Liquid Power

Urban and Industrial Environments
Series editor: Robert Gottlieb, Henry R. Luce Professor of Urban and Environmental Policy, Occidental College

For a complete list of books published in this series, please see the back of the book.

Liquid Power

Water and Contested Modernities in Spain, 1898–2010

Erik Swyngedouw

The MIT Press
Cambridge, Massachusetts
London, England

© 2015 Massachusetts Institute of Technology

All rights reserved. No part of this book may be reproduced in any form by any electronic or mechanical means (including photocopying, recording, or information storage and retrieval) without permission in writing from the publisher.

MIT Press books may be purchased at special quantity discounts for business or sales promotional use. For information, please email special_sales@mitpress.mit.edu.

This book was set in Stone Sans and Stone Serif by Toppan Best-set Premedia Limited. Printed and bound in the United States of America.

Library of Congress Cataloging-in-Publication Data is available.
ISBN: 978-0-262-02903-2 (hardcover)

10 9 8 7 6 5 4 3 2 1

To Maria, for everything

Contents

Preface ix
List of Acronyms xiii
1 "Not a Drop of Water . . .": Spain's Cyborg Water World 1
2 The Hydro-Social Cycle and the Making of Cyborg Worlds 19
3 "*Regeneracionismo*" and the Emergence of Hydraulic Modernization, 1898–1930 39
4 Chronicle of a Death Foretold: The Failure of Early Twentieth-Century Hydraulic Modernization 67
5 Paco El Rana's Wet Dream for Spain 99
6 Welcome Mr. Marshall! 129
7 Marching Forward to the Past: From Hydro-Deadlock to Water and Modernity Reimagined 163
8 Mobilizing the Seas: Reassembling Hydro-Modernities 191
9 Politicizing Water, Politicizing Natures, Or . . . "Water Does Not Exist!" 223
Notes 231
References 249
Index 285

Preface

This book project commenced on a rainy Saturday morning in 1995 in my office in the School of Geography at Oxford University. That morning I was trying to finalize a proposal for the European Union's Marie Curie Fellowship program that permitted senior researchers to undertake work in what the EU's bureaucratic jargon then defined as Objective 5 regions. The latter were regions that performed economically far below the EU's average and would benefit from "knowledge transfer"' by "experienced" scholars. At the time, most of Spain's regions (with the exception of the Madrid region and Catalonia) fitted the bill. At the same time, Spain was also suffering from an extraordinary period of drought that had lasted several years and had sparked an intensely politicized debate over water and water resource management.

For these reasons, it offered an appealing laboratory to extend into new domains my political-ecological work on the relationships between social and political power on the one hand and water and its hydro-social organization on the other. The problem was that I needed a personal invitation from a local university. So, I picked up the phone to call my old friend and former co-fellow at the Centre for Metropolitan Research and Planning at Johns Hopkins University, Fernando Molini, now professor of geography at Madrid's Autonomous University. I had not seen him for over ten years and I was not even sure if the old phone number I had in my address book was still correct. But as soon as the receiver at the other end was picked up, I recognized Fernando's voice.

I explained my predicament and he immediately offered to look into it. Just a few hours later I received a telephone call from Professor Immaculada Calavaca, head of the geography department of the University of Seville. By the next morning, all the paper work was sorted and I was set for my first foray into researching Spain's tumultuous hydro-social landscape. Upon arrival in Seville, I met Professor Leandro del Moral, the great leading national expert on water politics, who was to become my mentor and good friend. Over coffee in the university's bar, I explained my interest in Spain's waters and asked for some initial reading material. After finishing our coffee, Leandro took me through the corridors of the university, located in the old tobacco factory where Prosper Mérimée's Carmen, immortalized in George Bizet's opera, was rolling cigars in the nineteenth century. Upon unlocking the door to his office, Leandro said: "Here it is. Take it." Shelves stacked with literature and documents on Spain's waters lined the office walls. Without Leandro's generosity in sharing years of accumulated work, this book would not have been possible. Leandro, his wonderful wife Angela, daughter Lucia, and now grandson Nilo, have become great friends over the years, offering a welcoming home each time I return to this most beautiful and exuberant of Spanish cities. Indeed, Seville became a second home to me as the project took shape over the years following my first meeting with Leandro, and the conversations and discussions with academics, activists, and policymakers in Andalusia proved to be invaluable for shaping and sharpening the arguments that unfold in this book. Leandro's extensive network of contacts was also vital in helping me to navigate through the labyrinthine and highly contentious assemblages and constellations that shape and reshape Spain's hydro-social cycle.

A few years later, in 2004, I spent a sabbatical leave period at the Autonomous University of Barcelona. David Saurí from the geography department and Joan Martinez-Alier and his team at ICTA (Institut de Ciència i Tecnologia Ambientals) were great hosts. David is both an expert on Spain's water and a Tintin fan. No wonder we hit it off immediately. My successive research visits to Barcelona made me discover another one of Spain's exuberant cities and offered a rather different perspective on Spain's tormented twentieth century political-ecological transformations.

These two cities, where I would return many times, provided a welcoming base from which I ventured into the excavation and reconstruction of water's liquid power and the exploration of how Spain's waters are involved in shifting geometries of economic power networks, changing political alliances, competing cultural discourses, and technological controversies. I visited dams and irrigated fields, canals and desalination stations. I talked to local policymakers and representatives of regional or national government departments. I delved into archives and watched hours of film reels. I interviewed activists and academics, engineers and politicians, laypersons as well as experts. All of them I thank for their generosity and their willingness to share their knowledge and views with me.

Preface

The list of people who have been vital in bringing this project to fruition is indeed too long to recount and I cannot possibly do justice to all of them. Arantxa Rodriguez in Bilbao, Giorgos Kallis and his wonderful team of activist-researchers in Barcelona, and Antonio Rico and María Hernandez in Alicante were fantastic hosts and became dear friends.

I am of course also indebted to the two fine institutions where I worked over the past fifteen years, first at the School of Geography and the Environment of Oxford University and, since 2006, in Geography at the School of Environment, Education and Development of the University of Manchester. The latter, in particular, gave me valuable time and intellectual space, as well as an academic milieu conducive to debate and discussion with excellent students and scholars. These conversations too had their imprint on the project and helped shape the arguments developed in this book. I am particularly grateful to Gavin Bride, Stefan Bouzarovski, Saska Petrova, John O'Neil, Frank Moulaert, Simon Guy, Noel Castree, Vladimir Jankovic, Nik Heynen, Pedro Arojo, Neil Coe, Martin Evans, Esteban Castro, Mark Usher, Jason Beery, the Manchester OpenSpace crew (Brian Rosa, Joanna Tantanasi, Lazaros Karaliotas, Sampson Wong), Andy Merrifield, Manchester University's Society Environment Research Group, David Harvey, Robin de la Motte, Guy Baeten, Joseph Garí, Greet Remans, Stuart Franklin, Ben Page, Karen Bakker, Alex Loftus, Jessica Budds, Christine McCullough, Kerem Oktem, Morag Torrance, Michael Ekers, Miranda Morgan, Simon Addison, Suraya Fazel-Ellahi, Creighton Connolly, JulieAnn De Los Reyes, Santiago Gorostiza (especially for help with sourcing some of the images), and Melissa Garcia Lamarca. I am also grateful to Alba Vidal for her research assistance. The quality of the manuscript improved considerably thanks to the competent and meticulous proofreading by Hounaida Abi Haidar, Ben Lear, and Eva Swyngedouw.

Cartographers and graphics specialists Ailsa Allen from Oxford University and Nick Scarle from the Manchester University Cartographic Unit took care of the artwork, figures, and maps. It is great to work with such skilled specialists. Thanks also to Clay Morgan, Miranda Martin, and Kathleen Caruso at the MIT Press for their support, encouragement, patience, and professionalism.

Most of the research on which this book is based was funded by the EU's successive Framework Programs, most recently the ENTITLE Program, a Marie Curie Initial Training Grant, coordinated by the Autonomous University of Barcelona, and which brings together a fabulous bunch of junior and senior political ecologists from around the world. It is this kind of project that makes me believe that the European Union is still something worth fighting for, despite the widespread despair over the direction that this geopolitical project is currently taking. The British Academy funded part of the fieldwork on the politics of desalination. The bulk of the manuscript was completed during my tenure as a British Academy/Leverhulme Trust Senior Research Fellow in 2011.

I am particularly grateful to the National Archives of the Administration of the Spanish Ministry of Education, Culture and Sports (*Ministerio de Educación, Cultura y Deporte. Archivo General de la Administración*), the archives of the Junta de Castilla y León, and Antonio Rico of the geography department of the University of Alicante for their help in securing some of the images. Chapters 3 and 4 draw partly on earlier work published as "Modernity and Hibridity: Nature, *Regeneracionismo*, and the Production of the Spanish Waterscape, 1890–1930," *Annals of the Association of American Geographers* 89 (3) (1999): 443-465. Chapters 5 and 6 have been expanded and adapted from "TechnoNatural Revolutions—The Scalar Politics of Franco's Hydro-Social Dream for Spain, 1939–1975," *Transactions, Institute of British Geographers New Series* 32 (2007): 9–28, and from "Producing Nature, Scaling Environment: Water, Networks, and Territories in Fascist Spain," in *Leviathan Undone? Towards a Political Economy of Scale*, ed. R. Keil and R. Mahon (Vancouver: University of British Columbia Press, 2009), 121–139. A summary of chapter 8 was published as "Into the Sea: Desalination as a Hydro-Social Fix in Spain," *Annals of the American Association of Geographers* 103 (2) (2013): 261–270.

Over the many years it took to complete this book, I saw my children Eva, Nikolaas, and Arno grow up to become the wonderful adults they now are. They had to live through long absences and endure my endless conversations and discussions about water, politics, and emancipatory politics, but they also accompanied me during delightfully intense times in Spain. Last but not least, I dedicate this book to Maria Kaika, for everything.

List of Acronyms

AEDyR	Asociación Española de Desalación y Reutilización (The Spanish Desalination and Water Reuse Association)
CEH	Centro de Estudios Hidrográficos
CHS	confederaciones hidrográficas sindicales
CODA	Coordinadora de Organizaciones de Defensa Ambiental
DGOH	Dirección General de Obras Hidráulicas
EEC	European Economic Community
ETA	Euskadi Ta Askatasuna (Basque Homeland and Freedom)
EU	European Union
GATT	The General Agreement on Tariffs and Trade
HCSR	High Council for Scientific Research
INC	Instituto Nacional de Colonización
IPCC	Intergovernmental Panel on Climate Change
NATO	North Atlantic Treaty Organization
NHP	National Hydraulic Plan
NO-DO	Noticiario Documentales Cinematográficos
NWC	New Water Culture
OECD	Organisation for Economic Co-operation and Development
POH	Plan de Obras Hidráulicas
PP	Partido Popular (Conservative Party)
Programa A.G.U.A.	Programa Actuaciones para la Gestión y la Utilización del Agua

PSOE	Partido Socialista Obrero Español (Spanish Socialist Workers Party)
RBA	River Basin Authority
ROP	*Revista de Obras Públicas*
SIEHNA	Sistema Integrado de Equilibrio Hidráulico Nacional (Integrated System of National Hydraulic Equilibrium)
TVA	Tennessee Valley Authority
UAE	United Arab Emirates
UN	United Nations
UCD	Unión de Centro Democrático (Union of the Democratic Centre)
UNESCO	United Nations Educational, Scientific and Cultural Organization
WFD	(European) Water Framework Directive

1 "Not a Drop of Water . . .": Spain's Cyborg Water World

I am planning something geographical.
—Klaus Kinski, in Werner Herzog's *Fitzcarraldo*

Not a drop of water should reach the ocean without paying its obligatory tribute to the earth.
—Cortes Generales (Spanish parliament), 1912

This book explores how water becomes enrolled in the tumultuous process of modernization and development, and how the qualities and powers of water fuse with social, political, and economic processes in the pursuit of social dreams and fantasies nurtured by a diverse set of social actors. It is the stuff of visionaries and dreamers, bankers and builders, engineers and scientists, workers and peasants, states and industries, peoples and natures. Yet, the transformation of nature and society is also impregnated with vitriolic power struggles, entrenched territorial conflicts, overt or hidden processes of empowerment and disempowerment, of glory and defeat, of life and death.

Movies often capture this tumultuous process in ways that scholarly writings rarely succeed in doing. For example, Werner Herzog's iconic film *Fitzcarraldo* explores with fanatical zeal the disturbing, yet enthralling and captivating process of modernization as a torrential whirlpool that blends together individual and collective will to power, environmental transformations, phantasmagorical imaginaries,[1] and ruthless determination in a story of the harrowing passage of a few people in a ramshackle boat on the Amazon River in search of rubber and capital. Fitzcarraldo, the anti-hero of the movie, opera-lover and failed businessman, is set to bring civilization to a small town in the Amazon jungle. In a frenzied bid to mobilize the necessary capital to build an opera house in town—the great symbol of western modernity and civilization—he embarks on a desperate mission to unlock the potential of still-untapped rubber reserves deep in the rainforest. To do so, he needs to seek a passage—in this case a waterway—to transport the rubber from forest to market. The quest to find this mythical path through the rainforest turns into a heroic-tragic journey, riddled with courageous and unwavering determination to succeed in the face of an adversarial nature and society, ruthless mobilization and exploitation of indigenous people, and endless difficulties and obstacles. When his quest begins to unravel in abyssal turmoil and horror, and his few remaining friends and workers begin to abandon the doomed project, he shouts in despair from the top of the mountain, "*I am planning something geographical.*" The opera house will never be built, but the memory of the journey remains. Rarely has the project of modernity in all its horror and sublime beauty been so dramatically, yet clearly depicted as a deeply geographical affair: One articulated not only around the "conquest of nature" but also, more profoundly, around the will to transform nature, the making of a new socio-natural configuration, and with it, a new society, a new culture, a new environment, and a new polity.

This book is precisely about modernity as a geographical and environmental project or, more accurately, about how the production of new geographies and new "natures," both materially and symbolically, constituted both the basis of and condition for modernity, a process both sublime and horrific, emancipatory and oppressive, poetic and violent. Fitzcarraldo's shout in the jungle captures in its despairing passion the intellectual spirit that animates this book.

In Spain too, modernity unfolded as both drama and adventure, tragic and heroic, but above all, it progressed through the momentous and turbulent transformation of the environment. And it too combined shock and awe, despair and desire, violence and hope, failure and determination, exploding in all its contradictory forces around the turn of the twentieth century. The story told in this book is sandwiched between two emblematic moments. It begins in 1898, the year during which the loss of Spain's last overseas colonies triggered a period of cultural anxiety, and social and political turmoil still referred to as *El Desastre* (The Disaster). And it ends in 2010, the year that signaled the beginning of another period of deep crisis, cultural malaise, and intensified social and political conflict that profoundly shook the foundations of Spain's political-economic edifice. The book presents Spain's tumultuous history between these two emblematic moments as a political-ecological project marked by intense social and political conflict and brutal civil war and dictatorship, but also punctuated by periods of great hope and expectations for democratic reform, and sociospatial and cultural change.

When in 1898 the United States Navy fleet sunk the once proud Spanish Armada off the Cuban coast and terminated four centuries of Spanish global imperial rule, the aftershocks of this disaster were palpable in Spain for decades to come. The defeat confirmed the emergent mood of doom and moral decadence that had been brewing for decades among some of the Spanish cultural and intellectual elites as the empire crumbled. Military and imperial downfall mirrored the creeping realization that Spain was in deep economic, political, social, and cultural trouble. "The Disaster"—as the defeat is generally referred to—forced the Spanish old and new elites to shift their gaze to the national condition as the ultramarine horizons that had nurtured the imperialist project were now forever closed. The geographical shift of perspective to the motherland found a world not drenched in milk and honey, a fantasy and imaginary that the dominant imperial cultural elites had nurtured and celebrated for centuries. Instead, the rediscovery of inland Spain revealed unimaginable poverty in a countryside where widespread illiteracy coincided with despotic and feudal relations of power. Emergent capitalist industrialization in Catalonia and the Basque Country spurred mass migration and a budding labor movement. Impoverished peasants scraping by coexisted with a tiny class of large landowners. Social unrest intensified as the political landscape remained tightly controlled by the *caciques*, local political bosses who maintained, together with the military, a suffocating grip on power. A traditionalist Catholic establishment that instilled fear and religious zeal kept a tight control on both the souls and minds of their flock.

After *El Disastre* of 1898, national regeneration and the "rebirth of the nation" became the new rallying cry, one that would galvanize, mobilize, and animate modernizing political, cultural, and economic forces for much of the early twentieth century, while being resisted by a multitude of those who pursued different imaginaries for a postimperial Spain. The traditionalists yawned for a return to imperial glory while left revolutionaries conspired away to overthrow what they considered to be an ossified and hopelessly reactionary social and political order. In the interstices of these divergent visions, a new national modernizing imaginary gradually took shape.

The rediscovery of the inner condition of Spain, both culturally and physically, unleashed a feverish and highly contested process of transformation from an imperial to a postimperial and "modern" European state. Faced with competing states in the rest of Europe that had launched themselves on a trajectory of colonial modernization, Spain was forced to embark on what Michel Foucault would call a biopolitical national regime, centered on transforming people's lives (Foucault 2008). Regenerating Spain's land and people was staged as the only road open to redeem the country from its postimperial predicament and set in motion a process of a politically liberal, socially cohesive, and economically prosperous development. Modernizing Spain would prove to be a torturous historical-geographical experience, a process cut through by all manner of conflict, tension, and intense violence, but also marked by remarkable resilience and considerable successes.

Water played a decisive role in this process. During the twentieth century, the hydraulic configuration of Spain was completely overhauled. Today every single one of the ten mainland river basins is fully engineered, monitored, and manicured, and its waters are used for human—and increasingly "ecological"—purposes, often to the last drop. While throughout the twentieth century this process focused on engineering the flow of mainland waters, the twenty-first century's techno-natural configuration extends the engineering of Spain's hydro-landscapes to the Mediterranean waters by means of large-scale desalination projects. Dams, inter-river basin transfers, large-scale irrigation coupled with massive hydroelectrical development transformed the hydro-social cycle in radical ways, thereby producing new socio-ecological configurations in a hybridized landscape that fuses together things natural and social. Over the last few years, the quest for water has begun to capture the sea as large desalination facilities complement the terrestrial hydro-social cycle to assure quenching the insatiable thirst for water.

During the twentieth century, the social and political conditions indeed changed dramatically. From the intensely conflict-ridden early decades of the twentieth century, followed by the Franco era—one of the longest fascist dictatorships Europe has known—Spain became a liberal democracy in the final quarter of the twentieth century and decidedly embraced a trajectory of neoliberal global integration. The twisting and turnings of the hydro-social cycle both testify to and were active agents that animated the choreography of these seismic social and political transformations. Not only local and national relations played a decisive role in staging the drama of hydro-modernization, but also international geopolitical connections and strategies proved to be vital elements in shaping and reshaping the socio-environmental fabric of the hydro-social cycle.

The contested political-ecological process that marked the transformation of Spain's hydro-social landscapes during the twentieth century and into the new century will be the emblematic entry through which a wider set of issues related to nature, environment, modernity, and sociopolitical power are explored throughout the book. Its core focus is the social and material archeology of the making of the waterscape by means of assembling hydraulic, social, cultural, and political processes in a hydro-social constellation that shapes how, where, and to what water flows and how this circulation, in turn, becomes the arena of subsequent transformations. The stories narrated in this book aspire to move beyond H_2O, the stuff that comes out of taps and irrigation spigots. Instead, the limelight will be on the often-invisible actors and agents that are assembled in and through the flows of water and their benign or erratic behavior.

A Few Watery Vignettes

Paella is of course one of the classic delicacies closely associated with Spain's celebrated cuisine. The saffron-scented rice combined with shellfish, chicken, chorizo sausage, rabbit, or cuttlefish remains one of the culinary highlights of Mediterranean Spain. The Ebro Delta and along the lower stretches of the Guadalquivir River in Andalucía are among Europe's greatest rice-producing regions. Large-scale, intensive cultivation of cotton as well as tomatoes for industrial food processing add to the characteristic landscape of Southern Andalucía. Needless to say, rice production requires large volumes of water. This is not evident, however, in a region suffering from irregular rainfall and recurrent droughts. Indeed, gigantic hydraulic interventions were and still are required to assure that irrigation waters keep flowing. The once malaria-infested and hostile landscape required extraordinary efforts to harness its waters in a context in which neither capital nor technology was readily available. Achieving such a Herculean project was no easy task. It required extraordinary capital investments and technical expertise, some of it provided by the United States, and a ready supply of a huge labor reserve. Perhaps your next paella will taste somewhat different if you know that part of the irrigation waters that flood the rice paddies in Andalucía flow through canals dug by former political prisoners: lefties, anarchists, gypsies, and assorted other undesirables that were incarcerated by the fascist dictator who ruled Spain from 1939 until 1975. And of course, transatlantic financial support and expert advice played their usual Cold War roles in financing the fascist modernizing project.

The Spanish southern Mediterranean coast is of course a tourist paradise. Sandy beaches, sunny skies, proliferating resort development and great food attract millions of sun worshippers to spend their holidays there.

These sun-drenched landscapes produced the standard Northern image of the "good" life in the Mediterranean where easygoing lifestyles blend seamlessly with around-the-clock partying. Little do the tourist crowds know that the very possibility of such spiraling development is dependent on the mobilization of huge quantities of water in a region that in the near past sorely lacked the necessary resources to sustain such growth. Some of the water that comes out of the taps and showerheads in the resort towns is transferred from several hundred kilometers away, diverted from the Tajo River in the North and pumped, piped, and channeled to the local Júcar and Segura river basins, on its way piercing mountains and crossing valleys. Indeed, urbanization in southern Spain depends very much on intricate systems of ducts, dams, and reservoirs that are organized over very long distances to quench the thirst of the land and the people of the South. This hydro-infrastructure also feeds the sprawling greenhouse landscapes that cover much of the land in southern parts of Spain. Cut flowers and ornamental plants for export, undocumented Ecuadorian immigrant labor, intensive horticulture, and technologically sophisticated agricultural production systems rely on an intricately engineered and minutely manicured water management. Taking a shower in Marbella, one of the more posh resort towns, is indeed not as straightforward a process as it might seem.

Around the turn of the millennium, massive protests escalated in Spain, mobilizing hundred thousands of people, over planned major water diversion schemes that would transfer up to 4,000 hm^3 (cubic hectometers) of water from the Ebro and Tajo rivers to the deficit regions of the South and Barcelona. If implemented, the plan would radically reshape the hydro-social geography of Spain and require capital investments that would greatly outstrip those of the largest infrastructure project undertaken in Europe, the Channel Tunnel between France and the United Kingdom. While northern regionalists, ecologists, farmers, and assorted others opposed the plans, southern irrigators and development boosters demanded with equal zeal their "right to water" and to a "just distribution" of the nation's water. The ensuing Water Wars would intensify regionalist aspirations and fragment further an already precarious national geopolitical balance. Even Pope John Paul II intervened directly in the highly politicized debate when he declared with the Spanish Bishops that water should not be the province of some, but should be justly distributed for all.

The Book's Travel Guide

The preceding vignettes signal the importance of and active mobilizations of water. Rivers, seas, and aquifers, and the way they are enrolled and transformed, cast wide nets that link together a range of people in very different places, and express often contradictory, conflicting, and intensely controversial processes. Investment flows, political decisions, social struggles, geopolitical strategizing, the often-unexpected climatic and hydraulic vagaries, call for "water rights" and "water justice" for humans and nonhumans, and the ecological actions of various forms of water infuse and shape water fantasies, debates, and practices. Water imaginaries of all sorts become associated with regional cultures and specific water uses.

Water is indeed a deeply contested terrain, both theoretically and practically. While water flows through the hydro-social cycle, it connects all manner of bodies to technological systems, biological and chemical compounds, and each other. While water flows, it changes in character, content, and use. It flows through political jurisdictions and through organisms both human and nonhuman; it nurtures life and occasionally distributes death and destruction. Competing claims are made as it flows and different actors demand specific types of water at specific times and places. In this sense, water is intensely political in a very conventional sense as water is embedded in conflicting relationships of authority and power (Bakker 2012). But water is also political in a broader sense. With David Harvey, I contend that every political project is necessarily also an environmental project (Harvey 1996). This is one of the key themes that this book explores. Political dreams and aspirations, visions of development and change, ideals of community life, imaginaries of a "good" society—however diverse they may be—invariably rely on transforming environmental conditions and relations and imply the production of new socio-natural assemblages and constellations. This holds true for any process of social and political transformation. One of the tasks this book seeks to accomplish is to tease out how the diverse and contested political projects and plans that animated Spain's modernization process in the twentieth century were predicated upon particular trajectories of environmental transformation. More important, I shall argue, is that mobilizing particular imaginaries and projects of environmental transformation undergirded political programs and shaped state forms, institutional regimes, and geopolitical relations (Harris 2002; Kaika 2005; Bakker 2010).

Three interrelated theses inform the arguments developed in the book. First, modernization is explored as a heterogeneous and historical-geographically contested process that unfolds in and through a series of conflicting socio-natural transformations and revolutions. Second, this process is nurtured and sustained through the making and remaking of both physical and sociopolitical landscapes on the one hand and the reorganization of national and international scalar and geopolitical spatial relations on the other. Hydro-social systems and their political as well infrastructural production are mobilized as relationally constituted assemblages that stand as the symbolic and material emblem of modernization and geographical transformation, both locally and globally. Third, the book explores how the contested production of particular sociotechnical configurations (like dams, desalination technologies, irrigation systems, and inter-basin water transfers) depends on the assembling/enrolling of particular social groups, cultural discourses, technical expertise, material conditions, and political-economic power relations.

In sum, the book explores the interconnections of modernization, socio-environmental change, and the choreographies of social and political power. Water and hydraulic infrastructures play a lead role in shaping and organizing this process. The contested political-ecological process that marked the transformation of Spain's hydro-social landscapes during the twentieth century and into the new century will be the leitmotiv through which a wider set of issues will be explored. Spanish modernization was and is a decidedly geographical project and is expressed in and through the intense socio-environmental transformation of the country, both internally and in terms of its wider geopolitical relations. This transformation is one in which water and the waterscape play pivotal roles.

The book is theoretically and methodologically informed by contributions from science and technology studies, in particular perspectives that have contributed to understanding worldly things as assemblages of human and nonhuman relations and processes. In particular, Dona Haraway's cyborg metaphor and Bruno Latour's theorization of quasi-objects— things that are both social and physical—have been influential in shaping the argument (Haraway 1991, 1997; Latour 1993). Political-ecological perspectives, in particular the pioneering work of a range of scholars that took seriously the need to incorporate nature more centrally into historical materialist analysis, offered the theoretical framing that helped shape the analytical narrative of the book (Smith 1984; Harvey 1996; Swyngedouw 2004a; Gandy 2005; Kaika 2005). Finally, the seminal work of a generation of new environmental historians, in particular William Cronon's magisterial study of the urbanization of nature in Chicago (Cronon 1991), Richard White's moving story of the remaking of the Columbia River in *The Organic*

Machine (White 1995), and Donald Worster's unparalleled account of the mobilization of water in the arid U.S. West proved to be continuous sources of inspiration (Worster 1985). This book is indeed not just a story about Spain. A series of broader intellectual objectives inspired and informed the narrative that explores the transformation of Spain's hydro-social landscape between 1898 and 2010. First, I show how diverse political projects, social visions, ecological sensitivities, sociocultural imaginaries, discursive formations, institutional arrangements, economic interests and strategies, and engineering technologies fuse together in the making of particular environmental practices and the construction of hydro-technical infrastructures. Second, I document how human and nonhuman "actants" become enrolled in this historical-geographical process of multiscalar assembling. Third, I analyze the political-ecological processes through which particular sociotechnical configurations come into being and are stabilized, transformed, and ultimately replaced by other sociotechnical assemblages. And, finally, the book seeks to tease out the implications of this reading for contemporary environmental politics.

Spain's Torrential Waters

Spain is arguably the European country where the water crisis has become most acute in recent years. The political and ecological importance of water is not, however, only a recent development in Spain. Throughout the ages, water politics, economics, culture, and engineering have infused and embodied the myriad tensions and conflicts that drove and still drive Spanish society. Although the significance of water on the Iberian peninsula (see figure 1.1) has attracted considerable scholarly and other attention, the central role of water politics, water culture, and water engineering in shaping Spanish society on the one hand, and the contemporary water geography and ecology of Spain as the product of centuries of socio-ecological interaction on the other, have remained largely unexplored. Yet, very little, if anything, in today's Spanish political, social, economic, and ecological landscape can be understood without explicit reference to the changing position of water in the unfolding of Spanish society. The hybrid character of the water landscape, or hydroscape, comes to the fore in Spain in a clear and unambiguous manner. Every single one of Spain's ten mainland river basins is now fully engineered and manicured, often to the last drop of water. Every part of the terrestrial part of the hydrological cycle is subjected to some form of human intervention or use; not a single form of social change can be understood without simultaneously addressing and

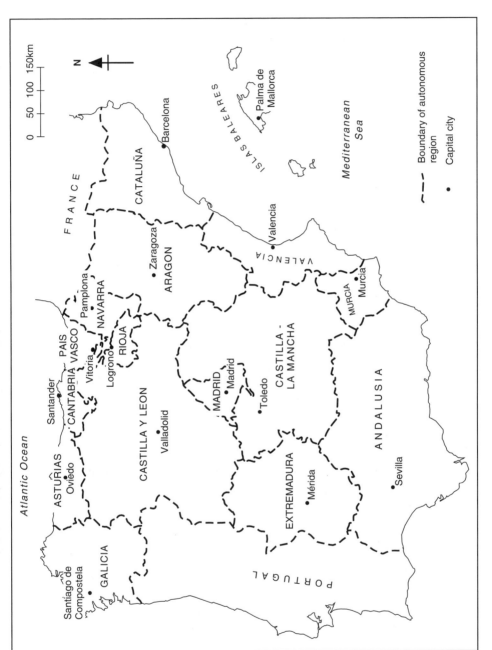

Figure 1.1

understanding the transformations of and in the hydrological process. The socio-natural production of Spanish society, I maintain, can be illustrated by excavating the central role of water politics and engineering in Spain's modernization process.

I shall not start by analyzing the Spanish water map from the available hydrological data and the physical characteristics of the water basins. Surely, such an entrée would be important, but prioritizing these would miss the central tenet of the premises on which this book rests. Indeed, these very physical conditions and characteristics are not absolute, stable, and God-given characteristics. On the contrary, the history of Spain's modernization has been a history of altering, redefining, and transforming these very physical characteristics. What is more, Spain's hydrological characteristics have been infused with social practices, cultural meanings, political and economic ideologies, and engineering principles for a very long time.

It was not until the late nineteenth century that the socio-natural production process of contemporary Spain accelerated. From that moment onward, Spain—belatedly, somewhat reluctantly, and almost desperately—launched itself on a path of accelerating modernization. Of course, modernization through socio-natural changes always takes place within already constructed historical socio-natural conditions. On occasion, I shall refer to and highlight these conditions. For the present purposes, which are to document and substantiate the notion of the production of nature on the one hand and to elaborate how Spain's modernization process became (and still is) a deeply geographical project, on the other, it will suffice to chart the tumultuous, contradictory, and often very complex historical geography of Spain's modernization through water engineering. Today, the country has over a thousand dams, more than eight hundred of which were constructed during the twentieth century (see figure 1.2). Water was indeed an obsessive theme in Spain's national life during the twentieth century—this theme and the quest for water continue unabated (del Moral Ituarte 1998, 2009).

My key objective is to grapple with the political-ecological dynamics that unleashed a dramatic socio-natural transformation articulated around the reengineering of Spain's terrestrial hydroscape. Figure 1.3 offers a bird's-eye view of what was achieved during the twentieth century, while indicating the distinct political-economic periods that animated the revolutionary production of a new hydro-geography. The figure summarizes the great expansion of dams and reservoirs, both in number and capacity between 1910 and 2000. Three distinct phases can be identified. During the first few decades of the new century, relatively little happened despite growing discursive and political attention to the urgent need for a profound hydraulic

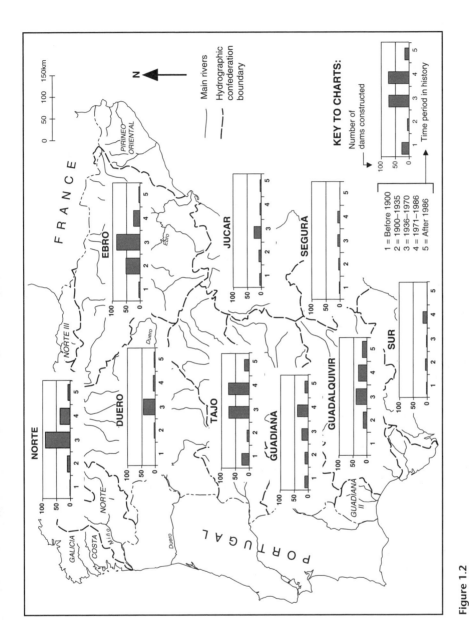

Figure 1.2
Evolution of dam construction in Spain for each of the hydrographic confederations (river basin authorities). Data exclude Islas Canarias and Pirineo Oriental.

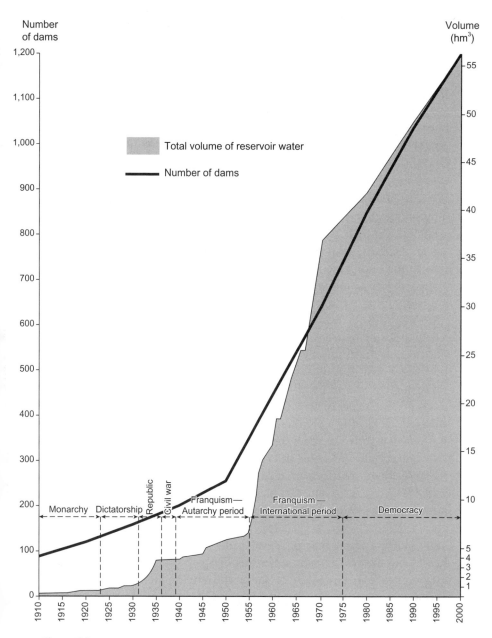

Figure 1.3
Evolution of dams constructed and volume of reservoir water in Spain, 1910–2000.
Sources: Author's elaboration, based on Diaz-Marta Pinilla [1969] 1997; Dirección General de Obras Hidráulicas 1990; Torán and Herreras 1977, 259–266; Martín Mendiluce 1996, 7–24; Ministerio de Medio Ambiente 2000b.

transformation. It was only during the dictatorship of Miguel Primo de Rivera and the subsequent Republic period that a surge in hydraulic interventions took place, to be interrupted by the period of the Civil War (1936–1939). During the fascist Franquist period and particularly after 1955, a real techno-natural revolution of Spain's waterscape unfolded. Indeed, General Franco realized his "wet dream" for Spain. There are clearly two periods in the making of this postwar fascist hydro-social landscape. The first period, between 1939 and 1955, was characterized by a sustained rhetoric of the urgent need for expanding irrigation and hydroelectrical production through the construction of state-led grand hydraulic works but with few real achievements. The acceleration of the scalar remaking of Spain as a hydro-social network had to wait until repositioning of the geopolitical relations, and their associated political-economic networking and flows of capital, expertise, and steel, took a radical turn after 1953, announcing a new period that proved to be a "watershed" moment in the realization of Franco's hydro-vision for Spain. Between 1955 and the end of Franco's dictatorship, most of Spain's terrestrial hydro-infrastructure was put in place. After the death of Franco and the transition to liberal democracy, state-led hydro-structural development continued its exponential expansion, only to slow down slightly by the end of the century. The inauguration of the twenty-first century was marked by paradigmatic shift in the political-ecology of Spain's hydro-structuralism as new dam and water transfer projects were increasingly contested and new social and political alliances emerged, mobilizing around radically new water imaginaries and practices. It is precisely the naked data summarized in figure 1.3 that will be ecologically and socially embodied in this book in an attempt to make water and hydro-technical transformation come alive and be politically performative.

Structuring the Hydro-Social Narrative

The following seven chapters develop the occasionally tragic, sometimes heroic, but always dramatic unfolding of Spain's modernization through a combined transformation of social and environmental relations. The perspectives deployed in the presentation of the historical-geographical transformations are explored in chapter 2. It presents a range of theoretical arguments and conceptual heuristic tools that have played a central role in weaving the storylines together. Nonetheless, this is not a theoretical book, but the more theoretically minded reader might find chapter 2 helpful in situating the lines of thought that infused the making of the argument. I mobilize a decidedly relational perspective, one that stitches together the physical and the social, the material and the symbolic, the local and the global in ways that elucidate how diverse instances, practices, things, and people fold into networks and construct more or less stable assemblages of socio-natural relations that become expressed in concrete things (like dams, canals, or desalination plants), organizations and institutions (like River Basin Institutions), practices and routines (like irrigation practices, water-use habits), visions and dreams. Chapter 2 explores from a decidedly historical and materialist perspective how such apparently disparate things and processes are imagined, fused together, maintained or transformed.

The six empirical chapters are broadly organized in chronological order. Starting in the late nineteenth century in chapter 3, the story reaches the early twenty-first century at the end of chapter 8. While the historical chronology provides an ordering over time, the conceptual emphasis nonetheless shifts and the relative importance of themes and practices changes as the narrative unfolds. The six empirical chapters cover three distinct periods, each of which foregrounds a particular set of theoretical and political-ecological concerns. Chapters 3 and 4 focus on the formation of a new radical imaginary that will frame the modernizing desires for most of the twentieth century. The emphasis here is on the contested construction of a new radical imaginary for Spain's modernization after the imperial defeat of 1898. Along with Maria Kaika's insistence on the importance of the imaginary in shaping the performative powers of a politics that envisages a transformation of socio-spatial configurations (Castoriadis 1998; Kaika 2010), the chapter considers the extraordinary reimaginations of Spain, modernity, and the role of water, spearheaded by an amorphous social group of emergent and aspiring elites who began to suture the cultural landscape and political discourse after the loss of the empire. The retrenchment of the elite's gaze to the homeland after the imperial adventure came to a close forced the gradual and deeply contested emergence of a new national myth, a new sense of identity, and the discursive production of an appropriate national polity of modernization. Chapter 4 switches from a focus on the imaginary to the truncated and ultimately failing attempt to launch Spain onto a path of hydro-modernization and political-economic integration into the European heartland during the first two decades of the twentieth century. The key emphasis of these two chapters is on the contradictory dynamics of fusing a particular discourse and imaginary of modernity with the actual messy practices of transforming the socio-physical landscape: the real of everyday political-economic and social struggles and conflicts, and the resistance of traditional social forces that often stood in the way of modernizing drives. I shall show that it is precisely these tensions and contradictions that stalled the modernizing transformations and, along with intensifying social, political, and inter-regional tensions, produced an inflammable and highly conflicting social terrain that would eventually plunge Spain's society into civil war and internecine feud. This first part chronicles the tale of a failure, of times and places out of joint, of imaginaries and fantasies that remain unhinged and detached from the formation of a hegemonic political and social project, despite desperate attempts by some to forge a shared vision and practice. The failure of modernization, and the incoherent and uneven mosaic of successes and disasters, intensifies an already highly fragmented and internally conflicting society. The Civil War of 1936–1939 will bring all this to center stage.

Chapters 5 and 6 turn to the post–Civil War period, when the authoritarian rule of Generalissimo Francisco Franco and his allies would push through the modernizing visions and plans of the earlier reformers. Under his rule, more than six hundred dams were constructed, cities and industries watered, hydroelectricity expanded, and millions of hectares of land irrigated. Here, our gaze will open up again to the world. The extraordinary transformations of the socio-environmental complex and the accelerated hydro-modernization of Spain rested, first, on the skillful combination of assembling all manner of human and nonhuman agents in a national project and the ruthless repression and silencing of those who disagreed. Second, Franco's staunch anti-communism assured the geopolitical insertion of Spain in the U.S.-dominated Western Alliance as the Cold War and its associated strategic geopolitics began to choreograph a new geometry of international relations. With Franco's Atlantic political allegiance, American and other Western capital and expertise began to flow to Spain. The fecund articulation of Spanish national integration with strategic international insertion would bring together plans, institutions, capital, energy, concrete, steel, and labor in a revolutionary transformation of Spain. Indeed, as will be shown, a geopolitical project is also an environmental project. While chapter 5 documents the arduous task of bringing together often unlikely partners in a common project of national development that build on a discourse of national integration and self-reliant development, chapter 6 shifts geographical scale and shows how it was precisely the insertion of a nationalist Spain within the Atlantic alliances that would foster the takeoff of hydro-modernization. Indeed, American capital and expertise was fused with the new local elites in one of the most remarkable modernization processes Europe has ever seen.

The final part of the book, chapters 7 and 8, concentrates on the last quarter of the twentieth century. This is a period during which Spain tried to deal with the demons of the past and transform into a liberal, democratic, and globalizing state. In chapters 7 and 8 the focus is on the social power dynamics associated with socio-ecological and political-environmental transition and transformation in a context of growing inter-regional tensions and socio-environmental conflict on the one hand and neoliberal European and global integration on the other. The focus of chapter 7 is, on the one hand, the dialectic between path-dependent persistence of the proven recipes of constructing even more terrestrial hydro-megastructures pursued by the booster networks assembled around the hydro-modern edifice during the Franco years; and on the other hand, the call for radical change and transformations advocated by the liberated voices that had been repressed for so long. Indeed, the hydro-structural imbroglio of the fascist period continued during the early "democratic" period despite the transformation of the political configuration. At the same time, new alliances and new connectivities were forged that would begin to transform both the imaginary of water and the question of how to manage the hydro-social landscape in new ways. The dialectic between the continuation of the past and the inauguration of the new will be the leitmotiv of this chapter. The global context nonetheless had changed in remarkable ways. In particular, growing international competition, the encroaching commodification of everything, and the consolidation of market principles as the key organizing scheme for transforming, distributing, and allocating resources had to be brought in line with a greater sensitivity to ecological issues and pressures as well as with growing local cultural and political demands and aspiring regional autonomy. These "glocalizing" forces—the simultaneous process of upscaling and downscaling of political and economic power—had indeed begun to undermine the centrality of the national state as the pivotal agent in organizing socio-environmental change (Swyngedouw 1997), forcing a reorganization of water policies. Chapter 8, finally, seeks to tease out the heterogeneous and often conflicting assembling of interests around mobilizing the seas and large-scale desalination as a new socio-ecological "fix" for Spain. It considers how desalination and the networks of actors sustaining its realization mark the transition from a state hydro-structural framework to a decentralized, yet still very much state-led market environmentalist, water management paradigm. The combined process of Spain's insertion within the geoeconomic and geopolitical imaginary of a neoliberal, but ecologically more responsible, European Union and the embrace of an untrammeled economic globalization drive nudged the politics of water into new sociotechnical trajectories that would both attenuate local conflicts and sensitivities and nurture a more market-conformist hydro-social logic. The chapter charts and explores the socio-natural actors that gather around this new, arguably revolutionary, socio-technological paradigm. The focus here is on the socio-environmental and ecological arguments advanced and how they gel in a new but fragile, incoherent, and often contradictory assemblage through which the desalination edifice was constructed as the emblematic discursive and material construct that permitted mediating and managing escalating tension and conflict. Extending the hydro-social cycle into the seas, while fully commodifying the production of clean water, announces a new ecological trajectory of modernization. Indeed, the techno-social edifice around water was transformed radically to make sure that nothing really had to change. The concluding chapter 9 explores how the preceding arguments open up new avenues from thinking and acting in the context of the escalating water problems and conflicts that plague not only Spain but many other parts of the world as well.

Before we embark on this mission, the next chapter will briefly explore the theoretical and conceptual perspectives that have both inspired and guided our archeology of Spain's hydro-social past and present, and that charts the contours of its future.

2 The Hydro-Social Cycle and the Making of Cyborg Worlds

There is no such thing as either man or nature now, only a process that produces the one within the other and couples the machine together . . . the self and the non-self, outside and inside, no longer have any meaning whatsoever . . . the human essence of nature and the natural essence of man become one within the form of production or industry, just as they do within the life of man as a species . . . man and nature are not like two opposition terms confronting each other . . . rather they are one and the same essential reality, the producer-product.
—Gilles Deleuze and Félix Guattari, *Anti-Oedipus*

The Production of Natures: The Genealogy of a Narrative

Every historical rupture . . . changes retroactively the meaning of all tradition, restructures the narration of the past, makes it readable in another, new way.
—Slavoj Žižek, *The Sublime Object of Ideology*

Taking Spain's postimperial condition as its entry point, the book seeks to elucidate the relationship between the hydro-social cycle—the socially embedded techno-institutional organization of the material flows of water—and the process of modernization as it unfolded during the twentieth century. The aim is to tease out the multiple relations of power through which "water" becomes enrolled, transformed, and distributed. Water is considered here as a hybrid flowing "thing" that fuses together physical, biological, social, political, economic, and cultural processes (Swyngedouw 1996). In doing so, I excavate how the materialities of nature enter the domain of the political and, through this, how environmental transformation can be read as a process of continuous assembling, disassembling, and reassembling of socio-natural relations. The primary objective is to demonstrate how every political project embodies a process of socio-environmental transformation and every socio-environmental project reflects and materializes a particular political vision. And it is precisely this thesis that renders water inherently political, and therefore contentious, and subject to all manner of tensions, conflicts, and social struggles over its appropriation, transformation, and distribution, with socio-ecologically unevenly partitioned consequences.

"Water" is indeed not just H_2O; its meanings and practices meander like rivers, making unexpected turns and gathering or assembling all manner of connections and relations, transforming the social and physical landscapes as it passes from source to sea. Despite its focus on the transformations of the hydro-social cycle in Spain, the ambition of the book is to broach a much wider range of themes and arguments that shine through the particular ways in which Spain's tumultuous hydro-social history unfolds. While both "water" and "Spain" figure centrally in the stories developed in the book, they are staged primarily as heuristic devices to help elucidate a series of theoretical and conceptual arguments.

Undertaking a socio-ecological archeology, the book reconstructs the process of the production of nature (Smith 1984, 1996), one that unfolds as a historical-geographical environmental narrative, that renders insightful the contingent and contested processes through which environments become constituted. Neil Smith's notion of "the production of nature," borrowed and reinterpreted from Henri Lefebvre, suggests that socionature itself is a specific and contingent historical-geographical process (Lefebvre 1991), insists on the inseparability of society and nature, and maintains the unity of socio-nature as a process. Smith insists that all natures, including social nature, are the temporary outcome of a dynamic and complex historical-geographical process (see also Levins and Lewontin 1985; Lewontin and Levins 2007). He argues that the idea of some sort of pristine nature becomes increasingly problematic as new "natures" are produced across space and over time. It is this historical-geographical process that led Donna Haraway and Bruno Latour to argue that the number of hybrids and quasi-objects proliferates and multiplies (Haraway 1991; Latour 1993). The modern world is a cyborg one, filled with proliferating socio-natural imbroglios. Indeed, from the very beginning of human history, but accelerating as the modernization process intensified, the objects and subjects of daily life became increasingly more socio-natural.

The narrative of this book attempts to transcend the binary formations of nature and society and to develop a language that maintains the dialectical relational unity of the process of change as embodied in and expressed by hydro-social constellations and their transformation. This suggests a process by which the human and the nonhuman become knotted together in increasingly intimate manners. The hydro-social landscape is viewed as an assemblage of interwoven processes that are simultaneously human, nonhuman, material, discursive, mechanical, and organic, but ultimately driven by political forces and economic processes that aspire to turning nature into capital, a process that necessarily implies changing social relations to nature. The myriad of processes that support and maintain social life and sustain economic growth (water, energy, dams, food, computers, etc.) always combine society and nature in infinite ways; yet, these hybrid socio-natural "things" are full of contradictions, tensions, and conflicts.

If I were to capture some water in a cup and excavate the networks that brought it there, "I would pass with continuity from the local to the global, from the human to the non-human," Bruno Latour maintains (Latour 1993, 121). As the case of Spain will demonstrate, water's flows narrate many interrelated stories of social groups and classes and of powerful socio-ecological processes that produce social and ecological spaces of privilege and exclusion, participation and marginality, chemical, physical, and biological reactions and transformations, the global hydrological cycle and global warming, capital, technologies, and the strategies of dam builders, land developers, irrigators and water institutions, the knowledge of engineers, the passage from river to reservoir, and the geopolitical struggles between social groups, communities, regions, and nations. The rhizome of underground and surface water flows and the streams, pipes, machines, and canals that come together in water gushing from fountains, taps, and irrigation channels are a powerful metaphor for a deeply interconnected socio-nature.

Capturing socio-nature, therefore, requires constructing multiple narratives that relate material practices, representational visions, and symbolic expressions. Each of these moments embodies particular characteristics that internalize the dialectical relations defined by the other domains, but none of which can be reduced to or determined by the other (Lefebvre 1991). In short, Lefebvre's triad opens up an avenue for enquiry that insists on the materiality of each of the component elements, but whose content can be approached only via the excavation of the metabolism of their becoming in which the internal relations are the signifying and producing mechanisms. Put simply, Lefebvre insists on the ontological priority of process and flux that become interiorized in each of the moments (lived, perceived, conceived) of the production process, but always in a fleeting, dynamic, and transgressive manner. However, following the maze of socio-nature's networks—as Bruno Latour suggests we do—is not good enough if stripped from the process of their historical-geographical production (see Escobar 1999). Hybridization is a process of production, of becoming, and of perpetual transgression.

Figure 2.1 summarizes this argument. None of the entries are reducible to the other, yet their constitution arises from the multiple dialectical relations that swirl out from the production process itself. Consequently, the parts are always implicated in the constitution of the "thing" and are never outside the process of its making. In sum, this perspective is a process-based episteme in which nothing is ever fixed or, at best, fixity is the transient moment that can never be captured in its entirety as the flows perpetually destroy and create, combine and separate. This dialectical perspective also insists on the non-neutrality of relations in terms of both their operation and their outcome, thereby politicizing both processes and fluxes. It also sees distinct categories (nature, society, water, techno-natures, etc.) as the outcome of the infusion of material-discursive practices that are each time creatively destroyed in the very historical production of socio-natural assemblages.

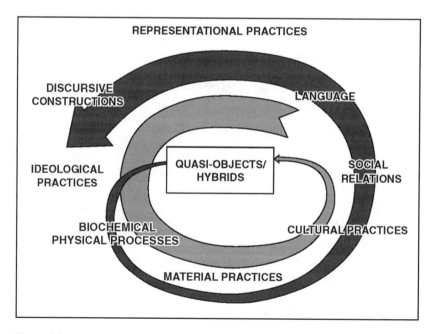

Figure 2.1
The production of socio-nature.

It is this perspective that David Harvey insists on as the epistemological entry into the excavation of the political-ecology of capitalism (Harvey 1996). A number of analytical tools arising from this formulation are useful for the political-ecological study of water and underpin the architecture of the stories narrated in this book. First, although we cannot escape the "thing," transformative knowledge about water and the waterscape can only be gauged from reconstructing its processes of production. Second, there is no "thing-like" ontological or essential foundation (society, nature, or text/discourse) as the process of becoming and of hybridization has ontological and epistemological priority. Third, as every quasi-object/cyborg/hybrid internalizes the multiple relations of its production, "anything" can be entered as the starting point for undertaking the archeology of her/his/its socio-natural metabolism. Fourth, this archeology has always already begun and is never ending and is therefore always open, contested, and contestable. Fifth, given the non-neutrality and intensely powerful forces through which socio-nature is produced, this perspective does not necessarily lead to a relativist position. Every archeology and associated narratives and practices are always implicated in and consequential to this very production process. Knowledge, discourse, and practice are invariably situated within the web of social power relations that defines and produces socio-nature. Sixth, the notion of socio-natural production transcends the binary distinctions between society/nature, material/ideological, and real/discursive.

While reconstructing the production processes of socio-natural networks along these lines is complex, I would maintain that such a perspective has profound implications for understanding the interconnections of capitalism, modernity, ecology, space and contemporary social life. It also deeply affects how one thinks about and acts on transformative and emancipatory ecological politics. Modernity itself, as a process, is internally deeply heterogeneous, contested and contestable, riddled with tension and conflict, and operates through and reorders society and nature in ever more complex and mediated ways (see Berman 1982; Harvey 1989; Giddens 1991; Braun and Castree 1998; Macnaghten and Urry 1998). This process relies on social, political, and economic processes as well as on shifting dynamics of incorporating all sorts of nature within the maelstrom of creative destruction that choreographs modernization. The latter is pushed forward by an unquenchable desire for growth, for accumulation for accumulation's sake, sustained by particular fantasies of what constitutes a good and just society, and predicated on the mobilization of all manner of natures in metabolic circulation processes through which nature becomes transformed, socialized, and distributed. Enrolling the fertilizing and modernizing powers of water was and is predicated upon enrolling all manner of things, actors, and processes in complex and heterogeneous, yet often unstable and invariably power-laden, assemblages that bring together human and nonhuman things in a relational configuration that extends from the local to the global. The expertise of engineers and scientists fuses with the hydraulic processes of catchment areas, rivers, and climatic dynamics; money needs to be turned into steel and concrete, and morphed into capital through enrolling all sorts of labor in a dynamic transformation process; imaginaries and fantasies of a green and abundant land, and a particular scripting of what nature is or can be needs to be coupled with engineering calculations, plans, institutions, governance arrangements, and legal codes. Landless peasants, smallholders and large landowners alike, workers and capitalists, political affiliations and visions of what is right or wrong, a wide range of institutional constellations, geopolitical tensions and alliances, the heterogeneous physical and ecological acting of water and all sorts of other nonhuman things and organisms, and much more, coalesce in the production of the hydroscape, the hydro-social assemblage that captures, transforms, metabolizes, moves, and distributes water to quench the thirst of cities, to fertilize the fields, and to permit all manner of other production processes to unfold.

Four key theoretical pillars provide the foundation for the narrative structure deployed throughout the book. First, a historical-geographical materialist perspective informs the presentation of the dynamics of socio-ecological transformation and accompanying changing social power geometries. Second, the "matter of matter" is taken seriously. I explore how changing views of what matter is and how matter acts become enrolled in new hydro-social practices. Third, the political-ecological metabolism of nature and the formation of socio-natural constellations are invariable articulated with territorially organized institutional and governmental arrangements. Attention will, therefore, be paid to how territorial scales of management—from local institutions to transnational governance—and social or political networking are made and remade in and through the contentious process that animates how particular socio-ecological configurations are inaugurated, shaped, stabilized, and transformed. Finally, the book is sensitive to the manner in which questions of socio-ecological or environmental justice and equality are articulated with discourses and practices of socio-ecological change. In the remainder of this chapter, I shall briefly outline the contours of these four theoretical perspectives.

One: Historical-Geographical Materialism

A historical-geographical materialist perspective takes the view that socio-ecological transformation unfolds through the combined social and physical transformation of the earth and is animated by the social relations through which access or ownership over nature, the socio-ecological modalities of its metabolic transformation, and the mechanisms of its distribution are organized. In the process, matter is morphed into "new" matter, while socio-ecological relations change and shift as the choreographies of social struggle and political conflict infuse and transform existing socio-ecological relations and new ones are created on the debris of older forms. Karl Marx insisted already in the nineteenth century that "the history of all hitherto existing society is the history of class struggles" (Marx and Engels [1848] 2012). Class struggle denotes a complex process whereby a seemingly coherent socio-ecological order is disrupted and contested, and struggles unfold over the content, direction, and organization of socio-ecological life. The modalities of enrolling nature—discursively, symbolically, and materially—into the relational fabric of life is the stake around which such struggles coalesce. Invariably, these struggles are waged primarily by the elites, the economically and politically dominant factions in society, in an attempt to maintain their privileges and material power while assuring a relative cohesive social order, but also denote the dynamics through which new social groups rise to dominance while others wither away. Although most interpretations of this perspective revolve around the choreographies of political and economic conflict, Marx's materialist interpretation insists that these power relations are invariably rooted in relations to nature and become performative precisely through these social relations to nature (Grundman 1991; Hughes 2000). In other words, social power and conflict unfold around the processes by which access to nature is socially organized, the way the metabolic transformation of nature is socio-ecologically structured and managed, and the mechanisms through which the results of this process are distributed. All manner of symbolic and discursive registers swirl around, shape the contours of, and inflect the trajectory and outcome of these struggles.

Under predominantly capitalist socio-ecological relations, these metabolic processes are inserted within the circulation of capital and driven by the incessant socio-ecological transformation of already existing "stuff" into new "stuff" on an ever-expanding scale and with an intensifying rhythm. The capitalist circulation of capital and its expansion, therefore, is of necessity predicated upon the socio-ecological circulation and metabolism of nonhuman matter whereby new and distinct socio-ecological configurations, in a material, political, and social sense, are constituted in the process. It does so through widening and deepening the socialization of nature and its incorporation within expanding metabolic processes.

Throughout the book, the emphasis will be on the dynamics of shifting class relations as expressed in new imaginaries and visions around water and their particular mobilization within institutional arrangements, changing state forms, and infrastructural constellations. In addition, the restructuring of class positions and the formation of networks of interest around the water edifice demonstrate how particular social groups are gathered around specific discursive formations, concrete state projects, particular technical infrastructures, and situated scientific knowledge. The historically highly variegated political-ecological landscapes that shaped the Spanish hydro-edifice demonstrate how Alan Schnaiberg's thesis about how the treadmill of production—the relentless search for growth, profit, and accumulation—provides an underlying rationale around which these diverse heterogeneous and changing modernizing projects and visions coalesce (Schnaiberg 1980). Indeed, the relative stability and coherence of socio-ecological relations is predicated upon their continuous and restless transformation.

The movement and transformation of matter, animated by specific social relations, can be captured through the notions of "metabolism" and "circulation." These are the central metaphors and material processes that will guide the endeavor to excavate the historical-geographical production of the hydro-social landscape in Spain (Swyngedouw 2006b). The changing metabolic circulation of water is presented as synecdoche for the changing political-ecological relations that animate Spain's modernization process. Engaging "metabolism" and "circulation" permit excavating socio-environmental constellations and their change over time. Marx and Engels were among the first to use the term "metabolism" to grapple with the dynamics of socio-environmental change and evolution (Fischer-Kowalski 1998, 2003). In fact, metabolism is the central metaphor for Marx's definition of labor and for analyzing the relationship between human and nature: "Labour is, first of all, a process between man and nature, a process by which man, through his own actions, mediates, regulates, and controls the *metabolism* between himself and nature. . . . Through this movement he acts upon external nature and changes it, and in this way he simultaneously changes his own nature" (Marx 1971, 283, 290; my emphasis).

For Marx, this socio-natural metabolism is the foundation of and possibility for history, a socio-environmental history through which the natures of humans and nonhumans alike are transformed (see also Godelier 1986).

For historical-geographical materialism, then, ecology is not so much a question of values, morality, or ethics, but rather a mode of "understanding the evolving material interrelations (what Marx called 'metabolic relations') between human beings and nature.... From a consistent materialist standpoint, the question is ... one of coevolution" (Foster 2000, 10–11; see also Norgaard 1994).

"Metabolism" is mobilized here in an ontological manner in which human beings, like society, are integral yet particular and distinct parts of nature (Padovan 2000). The original German word for "metabolism" is *stoffwechsel*, which literally translates as "change of matter." This simultaneously implies circulation, exchange, and transformation of material elements, organized under distinct social and political arrangements. While matter moves, it becomes "enrolled" in associational networks that produce qualitative changes and qualitatively new assemblages. These newly produced "things" embody and reflect the processes of their making, but simultaneously differ radically from their constituent relational parts. While every metabolized thing embodies the complex processes and heterogeneous relations of its past making, it folds in its specific and material manner into new assemblages of metabolic transformation. These dynamic heterogeneous assemblages form an expanding circulatory process that, under conditions of generalized commodity production, takes on the form of circulation of commodities and the circulatory reverse flow of capital.

New socio-natural forms, including the transformation of labor into labor power, are continuously produced as moments and things in this metabolic process. Whether we consider the production of dams, the reengineering of rivers, the delivery of potable water, the desalination of seawater, the management of biodiversity hotspots, the transfiguration of DNA codes, the cloning of species, the recycling of computer components by children in the slums of the global South's mega-cities, the trading of CO_2 emissions after Kyoto, the cultivation of potatoes (genetically modified or not), or the construction of a skyscraper, they all testify to the particular associational power relations through which socio-natural metabolisms are organized. Metabolic circulation is increasingly subject to the socially constituted dynamics of a capitalist market economy in which the alpha and omega is the desire to circulate money as capital (Douglas 1997). Yet, at the same time, this acceleration of movement and accumulation of capital also signals an intensified and accelerated accumulation of new natures, metabolized through metabolic vehicles that spin intricate networks and conduits (Virilio 1986). This leads to a proliferation of often unstable, invariably contested, and always temporary socio-natural assemblages whose destructive creation becomes one of the basic features of modernizing societies. All this, of course, raises the question of how precisely the matter of nature matters politically.

Two: Matter Matters Politically

The transformation of the social relies necessarily on a transformation of nature, of the nonhuman or, rather, more-than-human worlds we inhabit. As Jean Bennett argues, matter matters politically (Bennett 2010). Water is one of those emblematic forms of matter whose physical, chemical, biological, and cultural acting, as well as the forms of its sociotechnical mediation, scientific understandings, and discursive representations, matter significantly in political-ecological processes. The changing meanderings of water shape the particular trajectories through which it becomes engaged and enrolled in processes of hydro-social transformation (Linton 2010). The multiple characteristics and variegated acting of water, together with contentious demands, imaginaries, and uses are invariably mediated through technological systems, scientific knowledge, political institutions, economic forces, social practices, and policy networks and regimes. While this perspective implies a constructionist view in both a material and discursive sense, it does not suggest a smooth, linear, fully controlled, and thoroughly socialized process. The social never sutures the physical completely. As the arguments unfolding in the book will show, the construction of hydro-social assemblages that aim at socializing water in particular ways is fraught with conflicts and tensions and is filled by a wide and heterogeneous range of actors and agents, both human and nonhuman. Particular attention will be paid to how the variegated physical characteristics of water are staged and symbolized, and how particular qualities are signaled (while others are silenced or ignored). In doing so, I attempt to demonstrate that nature matters politically, but its politicized acting becomes performative through kindling specific relational configurations that attempt to foreground and enroll particular real and imagined qualities of water. Such an approach revolves around the archeology of what qualities of matter are rendered subject to public concern and how and by whom public dispute over these concerns is articulated. This perspective is far removed from any deterministic interpretation of the role of nature in socioeconomic development; it is one that recognizes that matter matters but this mattering is invariably socially, politically, and culturally filtered.

Socializing matter is also a process that inevitably seems to fail somehow, that stubbornly refuses to exhaust what is deemed right or even possible.

The Hydro-Social Cycle and the Making of Cyborg Worlds

The socialization of water's ecophysical properties, its insertion into complex constellations of hydro-social techno-natures like dams, reservoirs, irrigation canals, water bureaucracies, and hydro-policies, always leaves a remainder, a surplus, an excess that evades incorporation, symbolization, or assembling. Cyborgs and hybrids are imperfect creatures and cannot be but so. The attempts to engineer, master, and control the hydro-social cycle, to keep "nature on a leash," are never complete, never fulfill fully their Promethean promise. The actants in hydro-social networks often act more than expected. They push beyond the bounds in which they are imagined to dwell, and behave in strange and often unpredicted, if not unpredictable, ways. There is always a remainder, a hard kernel that resists incorporation. They invariably produce excess or a surplus, over and beyond the acting that permitted the hydro-social imbroglios to stand. Dams fail, land gets flooded, water bodies are contaminated, rain fails to come, electricity lines break down, projects are contested or shelved, pipelines get clogged up or burst, and ecological relations become reconfigured and produce unexpected consequences. Consider, for example, how dams collapse as flash storms undermine the dam's capacity to resist, how floods or droughts destroy carefully manicured irrigation landscapes, how farmers tap groundwater sources unsustainably to avoid paying for irrigation water, how hurricanes sweep away homes and crops, how local communities protest the diversion of "their" water for other uses and users, how fish populations, coastal lines, and other ecological relations act differently as the flows of water are redirected or reorganized. Water is not just source of life; it distributes death and destruction too. Its acting invariably spirals out of the straitjacket that attempts to manicure its flow and harness its powers.

Furthermore, what matter does and how it acts changes as it becomes transformed through socio-metabolic processes, reimagined in new ways, or managed and channeled through new governance arrangements. The gaps and remainders, the excesses and failures, in turn, form the fragments and filaments that potentially become the stuff from which new imaginaries and projects, alternative discursive enchainment, and new forms of political, social, and economic organization are engendered. This is precisely what permit prefiguring new hydro-social assemblages. The material acting of water, I contend, is itself a historically and environmentally constituted process; its political performativity changes as the fabric of hydro-social constellations changes too. For example, when the water imaginary in Spain shifts its gaze from terrestrial organization to include a marine perspective and the associated prospects of large-scale desalination, seawater's acting becomes an integral part of the formation of a new techno-natural imbroglio.

These disparate processes trace the global geographic mappings of the hydro-social flow that produce hydroscapes as palimpsests of densely layered bodily, local, national, and global—but geographically depressingly uneven—socio-ecological and techno-natural processes (Swyngedouw 1996, 2006a). The multiple temporalities and interpenetrating circulations of water through the hydrological cycle, dams, and reservoirs, canalization and distribution networks of all kinds, treatment and pumping stations, the flows of investment capitals, and so on, illustrate the perpetual metabolism and circulation of water and offer clues to the human and nonhuman relations that produce and sustain the hydroscape.

Three: Politics and Hydro-Scales

The flow of H_2O relates all manner of things and subjects in a network, a rhizome, connecting the most intimate of bodily and socio-spatial relations. It inserts them in a political economy of local, regional, national, and international development and is part of a chain of multiscalar circulations of water, money, texts, and bodies (Rothenberg and Ulvaeus 2001). The third theme that runs through the argument focuses, therefore, on the mobilization of scalar strategies and arrangements, and views the production of nature as part of a process of producing scale (Swyngedouw 2003; Sneddon 2003; Norman, Bakker, and Cook 2012). I start from the view that scale is not ontologically given, but socio-environmentally mobilized in socio-spatial power struggles and this, in turn, gives rise to occasionally significant new scalar configurations, while others become less important. In other words, socio-natural relations have a scalar constitution in a double sense. First, scalar relational *networks* are forged that produce spatial geometries that are more or less long, more or less extensive. These networks assemble local or regional interests but also relate to the formation of national or transnational networks (Swyngedouw 2004b). Second, these relational scalar networks articulate with produced *territorial* configurations like river basin authorities, regional authorities, nation-states, and transnational organizations (Zimmerer 2000; Natter and Zierhofer 2002).

In much of the literature on the scales of hydro-social constellations, four theoretical and empirical entries dominate the debate. The first of these arguments revolves around the relationship between hydro-social organization, hydro-technical interventions, and nation-state formation. Ever since Karl Wittfogel's seminal book on *Oriental Despotism* and his careful and sophisticated, yet ultimately environmentally deterministic, analysis of the relationship between socio-ecological practices in arid and semi-arid environments on the one hand and the nature of the despotic hydro-state on the other, research has proliferated on the close articulation between state formation and state structure on the one hand and the nature of hydro-social interventions on the other (Wittfogel 1957; Molle, Mollinga, and Wester 2009). Consider, for example, how German or Italian modernization and nation building unfolded through the ruthless transformation of the socio-physical conditions and characteristics (Blackbourn 2006; Armiero 2011), a process that became particularly visible and sensible during the intensely nationalist fascist regimes (Armiero and Graf von Hardenberg 2013). Similarly, water politics and hydraulic techno-natures are understood as pivotal ingredients of the historical construction of the national geographies and identities of Australia, the United States, or Canada and closely articulated with forging national identity and state formation (Worster 1985; Opie 1998; Dunlap 1999). The discursive and material mobilization of water and nature in nation-state building have of course been central leitmotivs in both socialist and postcolonial states too. The examples of the Soviet Union, China, India, South Africa, Turkey, or Egypt, among others, have become iconic examples of how the transformation of the intimate articulation between nature and society shaped both modernization and nation-state formation (Gadgil and Ramachandra 1992; McCool 1994; Shapiro 2001; Mitchell 2002; Jacobs 2003; Whitehead, Jones, and Jones 2007;Warner 2008; Harris and Alatout 2010; Harris 2012). In the Spanish context too, national integration, state projects, and large-scale hydro-infrastructures were intimately intertwined throughout the twentieth century.

The second perspective on understanding scale centers on the role of transnational watersheds and flows in choreographing inter-state geopolitical conflicts, and has equally been a recurrent theme in much of the water literature (Conca 2005). Particularly in water-scarce regions with shared watersheds, inter-state rivalries and competition over access and distribution of water have been subject to intense debate, often leading to apocalyptic imaginaries of pending "water wars." Controversies between Israel and Palestine, the riparian states of the Nile basins and the Mexico-U.S. shared water resources are well-rehearsed examples of transboundary water-conflict analysis (Selby 2003; Delli Priscoli and Wolf 2009). The analysis of transnational networks and forms of hydro-social governance has become increasingly influential in understanding hydro-social regimes. While these studies demonstrate how water is indeed subject to inter-state conflict, they rarely articulate such geopolitical conflict with how water is managed, organized, and engineered at other scales.

The third perspective focuses on the river basin as privileged analytical and managerial entry. Indeed, water-related research has paid considerable attention in recent years to the vexing question of river basin organization and management. Ever since Franklin Roosevelt inaugurated the iconic Tennessee Valley Authority (TVA) in 1933 as a part of his New Deal development plan to combat the deep economic crisis of the 1930s (see Miller and Reidinger 1998; Ekbladh 2002), attention has been paid to the formation of river basin authorities and the political-ecological controversies around river basin governance arrangements as vehicles for inclusive and participatory modernization. The TVA would become the emblematic example of integrated hydro-territorial development and a model repeated in many places around the world (Molle 2009).

As I shall show, Spain proved to be in fact an earlier pioneer of integrated water basin management. The Spanish Ebro River Basin Authority was established full seven years before the TVA and had already achieved major successes before the latter's establishment. Nonetheless, it would take many decades before the managerial water literature as well as hydrological science began to focus more centrally on the hydrological arguments, policy framework, democratic content, and institutional organization of river basin authorities. This is of particular relevance for the European Union where the establishment of river basin governance arrangements is a compulsory ingredient of compliance with the Water Framework Directive, but also in other parts of the world, river basin management has been propelled to the top of the water policy agenda. The scale of the river basin and its hydro-political configuration is indeed a vitally important territorial form of water governance. Its contested formation and internal dynamics cannot, however, be abstracted from its position within the wider operation of the state and its relation to international water regimes and policies.

The fourth debate focuses on the analysis of the politics of large hydro-infrastructure planning and the formation of localized large sociotechnical systems, whereby emphasis is placed on the multiple agents and actors that animate the often contentious implementation of large infrastructures and their articulation with scientific knowledge, engineering expertise, and the codification of technical specifications. Borrowing from science and technology studies, these perspectives tend to focus on the micro-choreographies of social and political power and how their performative agency becomes embedded in and expressed by the technical modalities and qualities of such large (or small) sociotechnical configurations (see, for example, White and Wilbert 2009; Barry 2013). While indebted to actor network theory and often inspired by Foucaultian approaches to governance and the exercise of biopolitical power, these perspectives rarely succeed in relating their explanatory framework to the analysis of the influential role of hydro-social processes that operate at other geographical scales.

Each of these four debates offers a privileged theoretical entry and possible lens through which to view the genesis of the twenty-first-century Spanish waterscape and sustain a particular scripting and understanding of Spain's water histories and geographies. However, the ambition of the arguments presented in the following chapters is not to foreground a privileged and particular scale of analysis, but rather to explore and reconstruct the historically and geographically contingent process through which hydro-social transformations unfold geographically and how particular scales of governance and of action, and their interaction, are fought over, implemented, and change over time.

While each of the preceding scalar perspectives informs the structuring of the narrative deployed in this book, our focus will be on the contested formation and dynamic articulation between territorial forms of hydro-social governance on the one hand and the scalar configuration of relationally constituted networks of social and political-economic power on the other. Not a single perspective is staged as an a priori frame in which to force the narrative. On the contrary, the unfolding of the narrative itself constitutes the key theoretical contribution by insisting on how the various scalar forms of governance, networks of social and political actors, regional and national state formation, and sociotechnical systems are co-constituted. The formation of river basin authorities, for example, will be explored in conjunction with national political-ecological transformations and with changing international geopolitical dynamics. Such territorial forms of multiscalar hydro-social governance, we contend, have to be understood in conjunction with the socio-spatial formation and reorganization of human and nonhuman networks. In doing so, we provide a frame of analysis that is both sensitive to the specificities and contingent moments of the Spanish example while nonetheless offering a mode of analysis and conceptual framework that can be productively mobilized for reconstructing hydro-social transformations elsewhere.

Four: Hydro-Justice—Whose Waters?

One of the pivotal questions that animate hydro-social conflicts and mobilize all manner of social groups around the architecture of the hydro-social edifice center on these questions: Who or what has rights to nature?

Who or what has rights to nature's water? And, how do we—humans and nonhumans—organize the metabolic process of assembling physical and social processes such that an egalitarian, democratic, and just hydroscape becomes produced?

In *Republic*, part of Plato's discussion of justice unfolds as a conversation between Socrates and the sophist Thrasymachus, whereby the latter claims that what is just is doing what is in the interest of the stronger. From such perspective, justice is not only about fairness but also about what is seen to be just by those in power. Appeals to justice have galvanized all manner of political-ecological visions and projects, some in the interests of advancing socio-ecological equality, others pointing in the opposite direction. The debate around justice has recently taken a new turn as environmental change and socio-ecological processes are increasingly shown to have highly unequal distributional outcomes (Holifield, Porter, and Walker 2010; Walker 2011). Environmental justice emerged both as a normative concept and a social movement, sustained by newly confirmed insights into the socio-spatially highly uneven distribution of environmental "goods" and "bads" (Schlosberg 2007). Environmental justice became defined and understood as a question of Rawlsian distributional justice. The latter is choreographed and structured by the highly uneven political and economic power relations through which decisions over environmental distributional conditions are made and implemented. Emphasis is thereby put on the socially and economically uneven positions of participants in the decision-making machinery that allocates the distribution of environmental goods and bads, revealing that environmental goods are partitioned such that elites benefit most while environmental bads are decamped to the areas of the powerless and disenfranchised. While nominally accentuating questions of equality, the emphasis of environmental justice perspectives is clearly on foregrounding liberal notions of procedural justice as fairness.

Nonetheless, the environmental justice perspective tends to be symptomatically silent about the particular ways in which political forms of power interweave with the specific modalities through which nature becomes enrolled in processes of capital circulation and accumulation, and how concrete notions of justice become cemented in such formations (Cook and Swyngedouw 2012). Therefore, the ways in which specific content is assigned to considerations of justice and its articulation with hydro-social transformations will be a recurrent theme throughout the book. Indeed, questions of hydro-justice and equity have become pivotal concerns in recent water-related scholarship (see, for example, Whiteley, Ingram, and

Perry 2008; Boelens, Getsches, and Guevara-Gil 2010; Arroyo and Boelens 2013). In contrast to the prevalent argument in the environmental justice literature that focuses on questions of just distributions, attention in this book focuses on the manner in which questions and issues of social, territorial, and socio-ecological justice are mobilized as an integral part of the process of constructing signifying sequences that permit allying often very heterogeneous social groups and bridging diverse if not opposing ideological positions. We shall demonstrate how conceptions of justice were framed, formulated, and enrolled by enlightened modernizers in the beginning of the twentieth century and how concerns of hydro-social justice became one of the leading metaphors that held together the heterogeneous alliance upon which the autocratic Franquist political project was secured and solidified. In doing so, we suggest that the notion of justice can and is mobilized in the context of highly divergent, if not plainly opposing, political narratives and projects and can, under specific circumstances, become a formidable rhetorical and political dispositive to nurture political projects that are radically antithetical to democratizing or more socially equal political-ecological projects. As the final chapters of the book will demonstrate, the uses of hydro-social justice may actually actively contribute to the formation and consolidation of techno-managerial hydraulic discourses and practices that foster a depoliticizing trajectory, whereby the foundations of a democratic politics that is rooted in assumptions of political equality are replaced by a set of practices that solidify profoundly inegalitarian socio-ecological relations.

Conclusions: The Metabolic Circulation of the Hydro-Social Cycle

The production of the hydro-social cycle through metabolic circulation is necessarily a process of fusion, of the making of "heterogeneous assemblages," of constructing longer or shorter networks. Metabolic circulation is the socially mediated process of environmental (including technological) transformation and transconfiguration, through which all manner of "agents" are mobilized, attached, collectivized, and networked. These assemblages of humans and nonhumans, of dead labor and vibrant materials, permit excavating the socio-environmental constitution of how socio-ecological power relations change over time. Nature and society are in this way combined to form a political-ecological imbroglio, a cyborg that combines the powers of nature with those of class, gender, and other relations. In the process, a socio-ecological fabric is produced that privileges some and excludes many.

While social relations of power are mediated through their articulation with physical matter, nonhuman "actants" play an active role in socio-natural circulatory and metabolic processes. It is these circulatory conduits that knit often-distant places and ecosystems together and permit relating local processes with wider socio-metabolic flows, networks, configurations of governance, and political-ecological dynamics. Territorial questions of river basin management, regional autonomy, statehood, and geopolitical relations intersect with networked alliances of social groups and economic interests that occasionally overlap with territorial forms of organization, but often disrupt or transform existing scalar configurations.

These multiscalar socio-environmental metabolisms produce a series of both enabling and disabling social and environmental conditions. While environmental social and physical qualities may be enhanced in some places and for some humans and nonhumans, they often lead to a deterioration of social, physical, or ecological conditions (or combination thereof) and qualities elsewhere. Therefore, processes of metabolic change are never socially or ecologically neutral. This results in conditions under which particular trajectories of socio-environmental change undermine the stability or coherence of some social groups, places, or ecologies, while their sustainability elsewhere might be enhanced. In sum, the book examines how the production of hydro-social configurations reveals the inherently contradictory nature of the process of metabolic circulatory change and teases out the inevitable conflicts (or the displacements thereof) that infuse socio-environmental change.

Social power relations through whom metabolic circulatory processes take place are particularly important. It is these power relations through which human and nonhuman actors become enrolled, and the socio-natural networks carrying them that ultimately decide who will have access to or control over, and who will be excluded from access to or control over, resources or other components of the environment, and who or what will be positively or negatively enrolled in such metabolic imbroglios. The recognition of this political meaning of nature is essential if sustainability is to be combined with a just and empowering development—a development that returns the environment and the choices inscribed in its myriad possibilities to its citizens. In other words, socio-ecological metabolisms are inherently political processes and, consequently, constitute an integral part of any political project. Political visions are, therefore, necessarily also ecological visions; any political project must, of necessity, also be an environmental project (and vice versa).

This conceptual and theoretical framework informed and guided the narrative structure and analytical composition of the chapters that follow. In many ways, the narrative is conceived as a film reel, in which each still reflects the movements that went before and shapes the ones to come. It is a journey into the intimate relationships that bind humans and non-humans together, and the possibilities and constraints they find on their way. Let's dim the light. The story begins.

El Duero cruza el corazón de roble
de Iberia y de Castilla.

                      ¡Oh, tierra triste y noble,
la de los altos llanos y yermos y roquedas,
de campos sin arados, regatos ni arboledas;
decrépitas ciudades, caminos sin mesones,
. . .
Castilla miserable, ayer dominadora
envuelta en sus andrajos, desprecia cuanto ignora.[1]
(Machado 1912)

3 "*Regeneracionismo*" and the Emergence of Hydraulic Modernization, 1898–1930

Here is no water but only rock
Rock and no water and the sandy road
The road winding above among the mountains
Which are mountains of rock without water
If there were water we should stop and drink
Amongst the rock one cannot stop or think
Sweat is dry and feet are in the sand
If there were only water amongst the rock
Dead mountain mouth of carious teeth that cannot spit
Here one can neither stand nor lie nor sit
There is not even silence in the mountains
But dry sterile thunder without rain
There is not even solitude in the mountains
But red sullen faces sneer and snarl
From doors of mudcracked houses
If there were water.
—T. S. Eliot, *The Waste Land*

La promesa de resurrección reside en la naturaleza.[1]
—Javier Varela, "Un Profeta Político"

In this chapter and in chapter 4, I excavate the origins of Spain's early-twentieth-century modernization process (1898–1936) as expressed in debates and actions around Spain's internal socio-natural conditions in a context of inexorable imperial decline. The production of a new society and thus a new geography revolved very much around the hydrological condition, and included the making and implementation of a new hydraulic paradigm. In this part, I shall explore the multiple actors, processes, ideas, dreams, and practices that would lay the foundations for this new modernizing imaginary and vision. The conceptual framework presented in chapter

2 helps structure a narrative that weaves water through the networks of socio-natural relations in ways that permit a recasting of modernity as a deeply geographical, although by no means coherent, homogeneous, totalizing, or uncontested project. If the social and the natural cannot be severed, but are intertwined in perpetually changing ways in the production process both of society and of the physical environment, then the rather opaque idea of "the production of nature" may become clearer. In sum, I seek to document how the socio-natural is historically produced to generate a particular, but inherently dynamic, geographical configuration.

Spain's Postcolonial Shock

1898 is etched into the consciousness of every Spaniard as the year of *El Desastre* (The Disaster). When on May 1 in Cavite in the Philippines and July 3 in Santiago de Cuba the U.S. Navy sent the remnants of Spain's once invincible but now ill-equipped and underfunded Armada to the bottom of the ocean, the global empire that Spain once was sank with it (Carr 1995; Figuero and Santa Cecilia 1998; Fusi and Palafox 1989; Pérez 1999). What had started a few years earlier as an independence movement in Cuba ended with a humiliating defeat at the hands of a greatly superior American war machine. Spain surrendered unconditionally and had to sign away, at a conference in Paris in December of that year, the last bastions of its global empire by granting independence to Cuba and ceding Puerto Rico and the Philippines to the United States. With a few canon salvos, thousands of deaths, and the stroke of a pen, Spain's four glorious centuries of global geographical dominance, rivaled only by the British, sizzled out (Serrano 1984; Smith 1994). The geographical expanse of Spain had been crumbling for decades and with it, the position of those elites whose status and power were directly linked with the political, military, and economic relations of empire. The defeat sent tremors throughout Spain and magnified an already acute sense of pessimism, decline, and despair.

Throughout the nineteenth century, political turmoil and socioeconomic stagnation in Spain had marked the unmistakable signs of an old world in crisis. Ever since the liberal revolutionary period of 1868–1874 (*El Sexenio Revolucionario*), and the subsequent Restoration period, the country's social and political mood had systematically darkened. Political unrest intensified, economic conditions worsened, and peasant poverty was rampant.

Social conflicts spread as emergent class struggle accompanied budding capitalist industrialization in Catalonia and the Basque Country. The empire had provided an escape route for many Spaniards fleeing poverty and seeking a new life. Its closure accelerated internal migration, particularly to the urban centers of Madrid and the North (Casado de Otaola 2010), further intensifying social tensions while accelerating rural depopulation. The "narrow political oligarchy" (Harrison 2000a, 4) was in disarray and illiteracy rates among the poor were close to 70 percent. The political system, while nominally democratic, was characterized by widespread *caciquismo*.[2] The two main political parties, Conservatives and Liberals, basically alternated in government through rigged elections. The middle classes, in particular, breathed a fin-de-siècle atmosphere of cultural pessimism and national doom. The *Leyenda de Oro* (the legend of a Golden Spain), the colonial elite's mythical presentation of imperial Spain as the land of milk and honey that had galvanized seventeenth- and eighteenth-century cultural imaginaries, had made way for a sense of despair, of pessimism, and of pending catastrophe (Castillo-Puche 1998, 20).

In the spring and summer of 1898, the Spanish empire "on which the sun never set"[3] came to an end, turning Spain into the first post-colonial European state. A sense of humiliation and bewilderment, of real and perceived "backwardness" compared with the other European states, disoriented the elites and spurred mounting internal conflicts. This sentiment was paralleled by growing introspection and a desire to diagnose and remedy the malaise from the part of those who recognized the need to redirect, transform, and reimagine Spain in order to discover and build a new future, for a modern Spain that would be on par with the other leading European nations. I take 1898 to be the pivotal symbolic moment, one that launched a new political sequence that would shape the century to come. Its effects are still rippling through Spanish society. Enlightenment, modernity, progress, economic power, and political glory could not any longer be secured through geographical expansion, imperial conquest, and colonial robbery. The external spatial fix had snapped brutally. There was nothing left other than to turn the gaze toward the lands and people of Spain itself. A new scalar gestalt imposed itself, articulated around a new vision and practice, centered on the internal geographical condition and the remaking of the national environment. This national modernizing effort would prove to be a torturous and tormenting process, pitting region against region, capitalists against workers, women against men, communists against anarchists, royalists against republicans, fascists against democrats. It would rip the guts out of the country, leaving an untold number of victims in unnamed graves on its way. The process initiated at the turn of the century would also remake Spain's geography and produce a radically new socio-natural configuration.

While other European imperialist countries were consolidating their geographical expansion overseas and shaping an imperial/colonial modernist project at the end of the nineteenth century, the traditional Spanish elites found themselves in a highly traumatic condition with the loss of their last colonial possessions. While progress, power, and national development seemed to depend on mobilizing the resources of overseas colonial space, this route to progress was cut off for Spain. Faced with a mounting economic crisis, rising social tensions, and an antiquated and still largely feudal social order in Central and South Spain that was lamenting the military defeat of 1898, Spanish progressive elites were desperately searching for a way to revive or to "regenerate" the nation's social and economic base (Ortega Cantero 1995). While the old elites desperately tried to cling to the traditional ways, it was the middle classes of artisans, small farmers, and an emerging industrial bourgeoisie, bolstered by a cultural movement that rediscovered, aesthetically as well as sociologically, the doldrums of rural life as reflected in the barren landscapes of both land and soul, that began to gnaw at the political and social edifice of post-colonial Spain (Harrison 2000b). These aspiring new social factions looked to the rest of Europe as the example to mimic.

This drive to revive the nation's "spirit" became known as *el regeneracionismo* (Fusi and Palafox 1998) and refers as much to a political-economic desire for modernization and development as to a scientific, cultural, and aesthetic movement (Harrison and Hoyle 2000; Casado de Otaola 2010). Emerging from growing discontent from the 1870s onward, *el regeneracionismo* became associated with a loose group of intellectuals, technocratic modernizers—in particular the Corps of Engineers—enlightened politicians, journalists, and the rising strata of middle-class farmers and industrialists. As will be discussed further in this chapter, *el regeneracionismo* also became associated with a literary movement, which José Martínez Ruiz, commonly known as Azorín and a leading member, coined in a 1913 article as *La Generación del 1898*. This Generation of '98 was particularly concerned with reviving and modernizing Spain in the context of the twin dramas of internal disintegration combined with political immobility and the loss of external imperial power (Figuero and Santa Cecilia 1998). Regeneracionism quickly became an "obsessive theme in national life" (Fernández Almagro 1970, 239) and produced a rich, albeit ambiguous and by no means homogeneous, ferment from which Spain would launch itself onto a path of modernization. The traditional elites, through their parasitic dependence on colonial exploitation, had until then successfully prevented or at least stalled this process.

The various regenerationist tendencies at the time shared a number of views: a concern with the "decadence" of Spain, a desire to regenerate the national "spirit," and a search for a foundation from which to launch a national revival (see Mallada 1890; Macías Picavea [1899] 1977; Isern 1899; Costa [1900] 1981; Ayala-Carcedo and Driever 1998). The main targets for their fury were aimed at "the motley assortment of parasites who dominated Restoration society: corrupt politicians, sterile bureaucrats, pious bishops, pedantic academics, and incompetent generals whom the dissidents judged responsible for the nation's decline" (Harrison 2000b, 56). Leading regenerationist intellectual Ramiro de Maeztu, for example, ruthlessly unmasked the myth of a glorious Spain and graphically described its condition as that of a "nation of fat bishops, general-fools, predatory politicians, conspiring and illiterate; they do not want to be seen in those barren plains . . . where twelve million worms live an animal-like life, doubling their body, to plow the land with the plow that the Arabs imported"[4] (Valladares 1998).

The heterogeneous themes that run through the work of the regenerationists center on a sensitivity to the arid and austere cultural and physical landscape of Central and South Spain, the disintegrated character (*la invertebración*) of Spain[5] and the need to revive Spain physically and spiritually. They all shared "a preoccupation with the state of the nation" (Butt 1980, 150). In particular, Spain's grueling socio-physical landscape and arduous geography were recurring themes, ones that stood as both symptom and symbol for the national condition. Ramiro de Maeztu, in his 1899 book *Hacia otra España*, captured both the despair and drama of Spain's heartland while asserting how both trees and water constitute the true foundations for producing wealth: "And what does one encounter on the immense plateau that stretches from Jaén to Victoria, from León to Albacete, from Salamanca to Castile, from Badajoz to Teruel? . . . What is Castile today? Walk in any direction. What is Castile today? A horrible wasteland populated by people whose apparent characteristic is hatred of water and trees; the two sources of future wealth!"[6] (de Maeztu [1899] 1997, 169). Santiago Ramón y Cajal (1899, 119–120), eminent biologist, concurred in his assessment of the Spanish geographical condition characterized by the "poverty of our soil—a vast barren central plateau dotted with oases and bordered by some tiny strips of fertile land—and the harshness of an almost African sky."[7]

In sum, while imperial countries pursued strategies of external spatial solutions, Spain was forced to revolutionize its internal geography and to produce new geographical configurations if it was to keep up with its expansionist northern European rivals. Literally, a new geography, a new socio-physical configuration would have to be actively produced. This program of producing new space embodied physical, social, cultural, moral, political, scientific, technological, and aesthetic elements, fusing them around the dominant and almost hegemonic ideology of national development, revival, and progress.

Modernization as a Geographical Project: The Production of Nature

The dominant form of socioeconomic development in Spain, until the late nineteenth century, combined colonial trade with domestic farming. The latter was based on large-estate dryland culture by primarily southern and central *latifundistas* (large landowners) whose economic position depended on a protectionist economic stance and a desperately impoverished semi-feudal peasant agricultural workforce on the one hand and *minifundios* (smallholdings), often extremely small, in the Northern parts of the country, on the other hand. Agricultural production revolved around the classic Mediterranean triad of low-yielding olives, wheat, and wine. The effects of increasingly liberalized international trade after 1880 (Carr 1995), combined with the loss of the last Spanish colonies in 1898, led to disastrous socioeconomic conditions and rapidly rising social conflicts that intensified already sharp social tensions in the countryside (Garrabou 1975; Fontana 1975). Cheap wheat imports from the United States, Tsarist Russia, and the Ottoman Empire undermined the competitiveness of local extensive dry-land wheat production (Harrison 1973). In addition, the migration of *phylloxera* from the United States to Europe in the mid-nineteenth century, imported by over-zealous botanists who brought American vines to Europe and, with it, the *phylloxera* bug, had devastated local vineyards (Harrison 2000a). The socio-ecological triad of species migration, the political-ecology of imperialism, and the scientific-economic quest for collecting new species played havoc with domestic ecologies. Spanish vines and the peasants cultivating them were particularly hard hit. Moreover, the traditional agrarian elites were faced with the emergence of modernizing agricultural and industrial elites who began to challenge the political-economic and ideological dominance of the *latifundistas*, while the proletarianization process, combined with sharpening crisis conditions, intensified class struggle in both the city and countryside.

Budding industrialization, mainly focused in the textile center of Catalonia and in the expanding steel and ship-building industries of the Basque Country (Angoustures 1995), sharpened the city/countryside divide and accentuated even more already longstanding inter-regional conflicts that nurtured the emergent demands for greater regional autonomy. Containing and working through these tensions without revolutionary transformation necessitated a vision around which the modernizing social groups could ally via a project of national regeneration. This revitalization, which became formulated as a project of geographical restructuring, combined major environmental change, socioeconomic restructuring, and moral revival, all of which were linked together in a regenerationist ideological discourse. Steven Driever summarizes this as follows: "Spain was portrayed as part of a new world order in which Spaniards had to interpenetrate with their natural environment and geographical space in order not to perish as the international marketplace reordered the world through economic competition" (Driever 1998a, 37).

Indeed, While many of the participants in the regenerationist debate centered this revival around the mobilization of the "natural riches and resources" of the country (Gómez Mendoza 1992a, 233), *el regeneracionismo* also emphasized the intellectual and moral revitalization of the people and the associated need for educational and scientific "progress." Modernization was, therefore, primarily seen as a question of voluntarism, enlightened education, scientific enquiry, technological innovation, infrastructural development, Europeanization, and moral rectitude. As Rafael Pérez de la Dehesa observed, the regenerationists mobilized "a pragmatic and scientificized language" that appeared to be "politically neutral" (Pérez de la Dehesa 1966, 168). As such, it operated over and above the daily quibbles of what was considered the antiquated and hopelessly inefficient political system of Restoration Spain. Such nonpartisan politics would permit welding the social together in a more coherent, less conflicting and opposing manner by assuring progress for all.

In sum, *el regeneracionismo* centered on a rediscovery of a forgotten Spain, a physical and social space disavowed by the imperial elites, but one that nonetheless—if properly attended to—contained the seeds for its own rebirth. Indeed, as the great regenerationist intellectual Unamuno wrote in 1902, "The future of Spanish society awaits within our historical society, in the intra-history" as opposed to the disaster and decline associated with the disintegrating imperialist ultra-marine project (de Unanumo [1902] 1998, 166). Santos Casado de Otaola (2010, 58) summarized the geographical regenerationist diagnosis and modernizing prophilatic remedies most eloquently:

Spain was largely a great forgotten hinterland. Terrible areas of degeneration and poverty that waited to be redeemed were hidden it its interior spaces, but also untapped sources of wealth that modern technology could put to productive use, and wonders bequeathed by nature and history, whose inventory would permit returning them to the annals of national memory. As a true colonial enterprise, there was a place reserved for the geographical, scientific and statistical study of the territory, with its dual promise of cultural prestige and strategic information. It was also about seeking to facilitate the expansion of the presence of the state as an agent of civilization and control, extending to where its capillary network did not yet reach.[8]

One of the key protagonists articulating this revival-through-modernization was Joaquin Costa (1846–1911).[9] He was the most prominent and visible figure of the radical regenerationism advocated by modernizing intellectuals around the turn of the century. Born the son of a poor peasant from Aragon and a self-taught all-round intellectual and onetime politician, his influential and prolific writings covered politics, social reform, education, and agricultural and hydraulic policies, among many other themes. His relentless struggle against poverty and the socioeconomic and political disintegration of Spain propelled him to the forefront of the intellectual debates at the time (see Ortega 1975; Pérez 1999). This *Lion of Graus* as he is customarily known coined the term "hydraulic regenerationism" that would dominate the modernizing debates for decades. He became the symbolic figure around whom much of the regenerationist tendencies crystallized and whose influence in shaping Spain's water modernization cannot be underestimated (Tierno Galván 1961; Cheyne 1972; Ortí 1984). In 1892, he wrote that state-organized hydraulic politics should be a national objective "capable of reworking the geography of the Fatherland and of solving the complex agricultural and social problems" (Costa Martínez 1911, 90; Costa [1892] 1975, 88). He captured the mood of the time, diagnosed the core of the national problem in strictly post-imperial terms, and identified the foundations for a remedial therapy that would launch the country onto a path of national modernization. It was a mission formulated as a military-geographic project, which he summarized as follows (Costa [1900] 1981, 13):

The disgrace of Spain originated principally because of the absence in national consciousness of the vision that *the internal war* against drought, against the rugged character of the soil, the rigidity of the coasts, the intellectual backwardness of the people, the isolation from the European Centre, the absence of capital, was of greater importance than *the war against Cuban or Filipino separatism*; and because of not having been as alarmed by the former as by the latter, and because of not having made the same sacrifices that were made for the latter, and of not having committed—sad suicide—the same stream of gold to the engineers and scientists as to the admirals and generals.[10] (cited in Gómez Mendoza 1992a, 233; my emphasis)

In his 1901 book *Oligarquía y Caciquismo*, Costa argued that the remaking of the fatherland would require the urgent implementation of a "veritable chirurgical politics," one that demanded "an iron surgeon" (Costa [1901] 1998, 115). Such an iron surgeon would possess the same virtues as Plato's philosopher-king: someone with an intimate knowledge of the anatomy of the Spanish people, with an infinite compassion for the people, who senses the pulse of the nation, possesses the qualities of a hero, and is outraged about the injustices inflicted upon the wretched of Spain. The leader of the regeneration of the fatherland must be a superior man, an iron warrior who is an enlightened politician, cultured, and who governs the people in order to ameliorate their condition (see Giménez Pérez 2002).

Spain's "geographical problem" became the axis around which both the sociocultural and economic malaise was explained, and where the course of action resided. This national geographical project would revolve around the hydrological/agricultural nexus. Costa's ideas and views were first aired at his famous lecture to the *Congreso de Agriculturos* in Madrid on May 28, 1880. In an ironic oratory style, he presented the *Leyenda de Oro*, the dominant cultural myth of Spain as a land of unlimited abundance (Costa Martínez 1911, 1–2): "There is no climate as mild as our climate, no sky as favorable as our sky, no soil as fertile and abundant as Spain's soil; here, nature generously and effortlessly provides for the sustenance of man. . . . The other people would die of hunger if we did not offer them the leftovers of this splendid feast to which Nature has perpetually treated us."[11]

Instead of reveling in a fiction of paradise, of looking at the Spain of legends through rose-tinted spectacles, endlessly repeated by the romantic aestheticism of the eighteenth- and nineteenth-century cultural, military, and political elites, the reality of the Spain that actually existed at the dawn of the new century needed foregrounding, it demanded nothing less than the urgent need to produce a new geography, a new space. Roughly one hundred years before Neil Smith (1984) coined the iconoclastic term "Production of Nature," Costa already insisted on the need to literally produce a new geography if Spain were to be delivered from the doldrums it was in and set itself on a course of modernization on par with its European neighbors. We have to recognize, he insisted, that

> our climate is the worst, our soil [the] least fertile, our sky the most thankless and parsimonious, our life the most painful and difficult, our people the most starved and ragged, our language the poorest, our intellect among the least fertile, our participation in the common work of human progress is less than zero; that there is no land in Europe that resembles less a paradise than the Spanish one. . . . [I]f in other countries it is enough for man to help Nature, here more needs to be done; is it is necessary to *create* her.[12] (Costa Martínez 1911, 3; my emphasis)

At base, the hydraulic foundation necessitating *el regeneracionismo* resided in the uneven distribution of rainfall, and the torrential and intermittent nature of Spain's fluvial system, which was said to make the country "the ante-chamber of Africa" (de Reparez 1906). Joaquín Costa summarized this as follows:

> The central plains, and perhaps half of Spain, is one of the driest regions of the globe, after the deserts of Africa and Asia. There are provinces like Murcia where one barely sees a cloud all year. . . . The atmospheric currents of the Mediterranean and the Atlantic do not bring to the parched fields of the Peninsula all the water plants need to grow and bear fruit, but there are immense deposits in the crests and bowels of the mountains, and we can, with mathematical regularity, distribute them over the land, crisscrossing the country with a hydraulic arterial system that mitigates its heat and quenches its thirst.[13] (cited in Gil Olcina 2001, 10)

For Costa, fusing the production of a new geography with a revolutionization of the internal operation of the state would help mitigate social tensions and provide the basis for a pro-modernist and popular petty commodity production–based development process (Ortí 1994). This focus on a geographical project as the foundation for modernization permitted progressive elites to raise social problems like class struggle, economic decline, and mass unemployment as important issues without formulating them in class terms. This, in turn, enabled the formation of an initially weak, but gradually growing, coalition of reformist socialists, populists, industrialists, and enlightened agricultural elites into a hegemonic block with a modernist vision of Spain's future—an alliance aimed to defeat the traditionalists *and* keep revolutionary socialists and anarchists at bay. Although coalitions, objectives, and means would change over time, the geographical basis for modernization would remain the guiding principle for this hegemonic vision that would become the pivot of Spain's development until the end of the Franco regime.

The realization of such an ambitious project of mobilizing resources and educating the people demanded thorough geographical knowledge of a particular kind. For the regenerationists, geography had to be put at the service of the rebirth of Spain, themes that were hotly debated in the recently established (1876) and very influential *Sociedad Geográfica de Madrid*. An intimate and in-depth knowledge and understanding of the physical and social geographies of Spain was considered a vitally important foundation for formulating strategies with which to facilitate a modernizing geographical reconstruction of Spain. As Gómez Mendoza and Ortega Cantero (Gómez Mendoza and Ortega Cantero 1987, 80) argue, "The real patriotism is the bedrock of the regenerationist project and this patriotism flows from the exact knowledge of the geographical reality of the country."

This concern for a reengineered geography is voiced by a series of contemporaries of Costa. In particular, Lucas Mallada (1841–1921), an engineer, geologist, and palaeontologist, had already in the 1880s lamented the fate of Spain and the "causes of the poverty of Spain's soil" (Mallada 1882; see also Ayala-Carcedo and Driever 1998) and advocated a program of reworking Spain's natural resource basis. As a key collaborator in the Commission of the Geological Map of Spain, established in 1849, Mallada conducted extensive fieldwork across Spain and called for a deeper and more detailed knowledge of its geography. His influential writings[14] were widely published, among others by Costa, who as editor of the publications of the *Institucion Libre de Ensenanza*[15] (Free Institute of Education) was close to Mallada's vision and by the *Sociedad Geográfica* where Mallada presented his controversial ideas. For him, the causes of Spain's poverty were both physical and social. While Mallada lamented the dryness of the climate, the absence of rain, the unfavorable orographic and geological conditions, he was by no means an environmental determinist. On the contrary, it was precisely the combination of ignorance of the real geographical conditions of Spain (particularly on the part of the elites) and the historical-geographical production of a particular socio-natural configuration that shaped the dire state of Spain. The general scarcity of trees and the massive deforestation was the result of how, "with vandalist instincts and irrational selfishness . . . the preceding generations had extensively devastated the forests" (Mallada 1890, cited in Martín Martín 2003: 254). Ricardo Macías Picavea (Macías Picavea 1895, 346), classicist and geographer, intellectual and journalist, summarized the relationship between the need for a scientific geography and the regenerationist project as follows:

To rehabilitate us, it is imperative to start with the rehabilitation of our land: this is an essential and absolute condition. And in order to restore this geographical environment that is our fatherland, who would deny that the first thing to be done is to know it well and accurately? Here you have therefore the truly patriotic character with which the nurturing of Iberian geography must offer itself to the eyes of all Spaniards.[16]

The role and importance of exact-positivist geographical knowledge was not only rehearsed continuously in the subsequent decades, but also its message would spread widely and resonate well with the aspirations of the modernizing technocrats par excellence, the civil engineers (discussion to follow). In 1918, Rafael Altamira, another leading intellectual, wrote how the description of Spain's geography offers a great lesson in patriotism (Altamira 1923, 168–169), while Azorín (José Martínez Ruiz), a member of the "Generation of '98," concluded in 1916 that "the basis of patriotism is geography" (Azorín 1982, 512). Even by 1930, this vision was still the primary leitmotiv for the regenerationist agenda, which, in fact, would only materialize on a grand scale after the end of the Civil War in 1939. A newly published journal in 1930 recapitulated the great geographical mission of the modernizing agenda with the same vigor and passion: "There is nothing more urgent for our national reconstitution than a profound study of our geography and our soil. This will be the seed for the great political re-birth of Spain"[17] (NN 1930, 29–30).

This project to remake Spain's geography as an integral part of modernization combined a decidedly political strategy, a particular ideological vision, a call for a scientific-positivist understanding of the natural world, a scientific-technocratic engineering mission, and a popular base rooted in a traditional peasant rural culture. Plenty of evidence can be found for this in Costa's work and in that of his contemporaries (for a review, see Pérez de la Dehesa 1966; Tuñón de Lara 1971; Ortí 1976; Casado de Otaola 2010). The revolution in the state—but certainly not of the state—effected through a politics of spatial and environmental transformation would center on the defense of the small peasant producer-as-landowner, communal (state) control of water, educational enhancement, technical-scientific control, and the leap to power of an alliance of smallholders and the new bourgeoisie that hitherto had been largely marginalized by the aristocratic landowning elite and their associated administrators in the state apparatus.[18] At the same time, the focus on restoring, redistributing, or, in fact, expanding land ownership through "internal colonization," would foster growth and concentrate the efforts of an "organically" organized state that would bring together reformist intellectuals, some worker movements, and the nascent industrial bourgeoisie in a more or less coherent vision of reform against the traditionalists (Ortega 1975). The geographical project became, as such, the glue around which often-unlikely partners could coalesce, while excluding both the more radical, left-winged revolutionaries and the "radical" conservatives. Surely, the sublimation of the many tensions and conflicts within this loose alliance of reformists, when accomplished through a focus on reorganizing Spain's hydraulic geography, served the twin purpose of providing a discursive vehicle to ally hitherto excluded social groups without defining the problem purely in class or other adversarial social terms (see Nadal Reimat 1981). As Alfonso Ortí (Ortí 1976, 179) maintains:

Hydraulic politics, understood in a broad and symbolic sense as a process of accelerated transformation of agriculture from extensive and traditional into modern and intense, must constitute the fundamental vector of national politics. This must catalyze an agrarian reform, which would permit a balanced economic development and prevent the on-going process of proletarianization of the peasant masses, moderating social polarization and class struggle.[19]

This organic and anti-revolutionary (in social class terms) reformism in which the state would take center stage to organize the socio-spatial transformation would, after the failed attempts to initiate reform during the first few decades of the twentieth century, provide a substratum on which the later Falangist, organicist, and fascist ideology would thrive.

Water as the Lynchpin to Spain's Modernization Drive

"Los Pantanos o la Muerte!"[20]
—Regenerationist saying, cited in Pérez, *Historia de España*

If the "remaking" of Spain's geography became the great modernizing adagio, then water and hydrological engineering were its master tools. The study of geography centered on problems of fertility, both the lack of water and the infertility of the soil. In 1903, Costa wrote that "the greatest obstacle which prevents our country to improve production is the absence of humidity in the soil because of insufficient or absent rainfall" (cited in Ortega 1975, 37), while torrential winter and spring surges eroded the land and gushed water to the seas. "Rain rushing to the sea and taking part of the soil with it" was to be avoided at all cost, a parliamentary document of 1912 stated, repeating the already century-old claim (which would be heard again during recent periods of intense drought) that "not a single drop of water should reach the ocean without paying its obligatory tribute to the earth" (Gómez Mendoza and Ortega Cantero 1987, 1992, 174). Indeed, the dominant view at the time was that "Spain would never be rich as long as its rivers flowed into the sea" (Maluquer de Motes 1983, 96). The regenerationist rhetoric assigned great symbolic value to the customarily invoked image of the "mutilating loss of the soil of the Fatherland as a consequence of the nature of the pluvio-fluvial regime" (Gómez Mendoza 1992a, 240). In addition to the temporally and spatially erratic and problematic character of the "natural" hydrological dynamics, the modernization of industry, the mechanization of agriculture, and accelerating urbanization generated a real "water fever" (Vilar 1968; Maluquer de Motes 1983, 84).

Joaquín Costa's "hydrological solution" would offer the substratum for fostering growth, permitting social and land reforms, and economic development as well as cultural emancipation (Ortega 1975). His writings would be invoked time and time again by a wide variety of social groups to defend and legitimize national hydraulic programs and the policy of land reform through "internal colonial" settlements. The regenerationist project became formulated as a "hydrological correction of the national geographical problem" (Gómez Mendoza 1992a, 236). In Spain, the God-given, but unfavorable, natural conditions have to be corrected; God made a mistake in Spain and it was for man to rectify the defects written into the hydrofluvial landscape. For Costa, hydraulic politics sublimated the totality of the nation's economic program, not only for agriculture, but also for the whole of Spanish socioeconomic life. Hydraulic constructions for irrigation purposes, which would remedy the "inconsistencies of its [Spain's] rains," were regarded as a progressive alternative to the traditional policy of tariffs and import restrictions, which were supported by dry-land *latifundistas* (Torres Campos 1907). Hydraulic interventions stood as the material and symbolic kernel around which the possibility for a national rebirth was articulated. In an interview on February 15, 1903, Joaquin Costa put it thus: "'Hydraulic Politics' is a trope, a sort of synecdoche that expresses in a figure of speech the whole of economic politics that the Nation has to follow to redeem itself, a sublimated expression of the political-economy of the Nation"[21] (Hernández 1994, 14).

While these hydrological conditions figured as descriptors of Spain's geography, they did not necessarily determine its fate. The great modernizing drive of the revivalists therefore demanded not only an imitation and use of nature, but also its *creation*: "[increasing] the amount of fertile soil by making a hydraulic artery system cross the whole country—a national network of dams and irrigation channels" (Gómez Mendoza and Ortega Cantero 1992, 174).[22] In *El Problema Nacional*, Ricardo Macías Picavea, a leading regenerationist essayist and intellectual, summarizes the hydraulic mission as a necessary strategy for national development:

> There are countries which . . . can solely and exclusively become civilized with such a hydraulic policy, planned and developed by means of designated grand works. Spain is among them. . . . And the truth is that Spanish civilized agriculture finds itself strongly subjected to this inexorable dilemma: to have water or to perish. . . . Therefore, a hydraulic politics imposes itself; this requires changing all the national forces in the direction of this gigantic enterprise. . . . We have to dare to restore great lakes, create real interior seas of sweet water, multiply vast marshes, erect many great dams, and mine, exploit and withhold the drops of water that fall over the peninsula without returning, if possible, a single drop to the sea.[23] (Macías Picavea [1899] 1977, 318–320)

This patriotic mission that required the convergence of all national forces around the hydraulic program became the embodiment and representation of a collective myth of national development. This project

"*Regeneracionismo*" and the Emergence of Hydraulic Modernization, 1898–1930   53

was sustained and inspired by what Ortí called "a reformist geographical optimism," which substituted for the social and political pessimism of Spain's turn-of-the-century condition (Ortí 1984, 18). The hydraulic utopia of abundant waters for all would not only produce "ecological harmony," but also contribute to the formation of a socially harmonious order. The production of a new hydraulic geography would reconcile the ever-growing social tensions in the Spanish countryside; tensions that were taking acute class forms and resulted from the adverse and conflictive conditions of scarcity and inequality. According to Ortí, the symbolic power of this material intervention in the production of a new hydraulic geography to achieve "hydraulic regeneration" constituted "a mythical power, a collective illusion and the imagined reconciliation of diverse ideologies" (12). This specific form of regeneration served the productivist logic of the new liberal bourgeoisie that aspired to transform society and space according to the principles of capitalist profitability and aimed at Spain's integration into Europe's modernization process. This desire for economic revitalization was accompanied and inspired by "a cultural turn," an aesthetic as well as sociological rediscovery of the condition Spain was in. A new mode of expressing and symbolizing the faith of Spain would emerge around the turn of the century.

Humming to Nature's Tune: The Generation of '98

Es la tierra de Soria árida y fría.
Por las colinas y las sierras calvas,
verdes pradillos, cerros cenicientos
. . .
La tierra no revive, el campo sueña.
Al empezar abril está nevada.[24]
—Antonio Machado, from *Campos de Soria*

El hombre y la tierra se hacen mutuamente.[25]
—de Unamuno, *Obras Completas*

Indeed, the hydraulic regenerationism described previously coincided with an intellectual and professional critical regenerationism, symbolized by the literary *Generación de 98*, which rediscovered, both aesthetically and sociologically, the underdeveloped regions of arid Spain. While symbolic representations of Spain by the traditional elites still reveled in perpetuating the "Leyenda de Oro" (see Driever 1998a), which portrayed Spain as a prosperous, thriving, and successful country, a new generation of essayists, poets, novelists, and playwrights emerged from the late 1880s onward. They were tormented by what was seen as the decadence and decay of Spain, culminating in the disaster of 1898, and desperately tried to articulate a new vision of possible regeneration. Although they were by no means a homogeneous group in intellectual, political, or philosophical terms, they shared a desire for a positivist-scientific representation of the fate of Spain's arid or semi-arid regions and a soul-searching quest to transform and regenerate both "soul and body" (Pérez 1999) of the people (Figuero and Santa Cecilia 1998). Although normally associated with a small group of intellectuals like Ramiro de Maeztu, Pío Baroja, Miguel de Unamuno, Azorin (José Martínez Ruiz), and Antonio Machado, the influence of the Generation of 98 was felt much wider and was an integral part of the regenerationist debate and modernizing desires that brewed around the turn of the century (Angoustures 1995; Figuero 1998). They lamented the decadence of the political elites and their only perspective for future emancipation resided in a spiritual and political rebirth of the nation and in embracing hydraulic politics (Mainer 1972).

The "hydraulic desire" of the arid lands became indeed the leitmotiv of much of the regenerationist literature at the time. The fate of the drylands stood as the symbol of the decline and failure of Spain to modernize. For example, in the novel *La Tierra de Campos* (1896), Ricardo Macías Picavea describes the thirst of the land and the hydraulic condition of the drylands as follows (Macías Picavea 1896, 151): "Autumn had passed without a drop of water, to the point that the seeds had dried out. The winter rains did not come either. The northeastern winds, dry and icy, haven't stayed away a single day. And what a period of frost! Black, scorching, without a drop of water in the atmosphere, with temperatures of twelve and fifteen degrees below zero."[26]

As the enigmatic hero of the novel attested, only a hydraulic quest could revive the land: "without the prior solution of the vital problem of irrigation, no significant reforms will be possible in this country"[27] (Macías Picavea 1896, 151). The Generation of 98 often invoked the writings of Joaquin Costa or Lucas Mallada to express and represent both the landscape and the sociocultural conditions of dryland Spain. Unamuno, for example, refers to symbolic representations such as "the cruelty of the climate," "the somberness of the landscape," or "the biting and dry soul" to evoke Spain's rural conditions and the characteristics of the people. Ramiro de Maeztu describes his native Castilian landscape as "provinces depopulated like Russian steppes" or "a horrible wasteland inhabited by people whose characteristic quality is their hatred of water and trees."[28] Although superficially similar to a crude environmental determinism, their views were rather more inspired by a desire to lift Spain from the doldrums of enduring political malaise and persistent poverty of the masses and to recapture its lost position as a great modernizing nation among the other European states. As Ortega y Gasset, the great Spanish philosopher, would put it later, "The arid land that surrounds us is not a fatality over us, but a problem in front of us" (quoted in Hoyle 2000b, 31). Their lyrical description of the *mesetas* of Castilla, the melancholic accounts of the barren, gray, dry, infertile landscape, their venomous hatred of the traditional conservative elites, and their desire for a voluntarist and heroic leader, who is scientifically informed yet compassionate, to lead the people and the country to a bright, modern and European future, are the themes that connect these various authors. For them, the symbolic and mythological powers of water are mobilized as the basis for the hydraulic and spiritual revival of the country. Contact with water holds the promise of regeneration, of a new birth, while immersion in water fertilizes and strengthens the powers of life (Ortí 1984, 17). Against the dryness of the land and the misery and frustration of underdevelopment, stands the abundance of water that puts the hydraulic utopia of the regenerationist discourse at the center of the promise of a revival of the vital energies of the country and a future abundant paradise (del Moral Ituarte 1998). Landscape, nature, and political-economic or sociocultural processes fused together in a mutually constituted relationship (Hoyle 2000a) that required, if Spain were to modernize, a complete environmental transformation to redeem the plight of post-imperial Spain and set it on a course to modernity.

The typical hydraulic heroes of the 1890s novelists were apostolic figures whose voluntarist vision fought against the desperation and ignorance of the rural masses and the persistent dominance of the traditional rural elites, imposing on their modernizing program a hydraulic revival meant to resolve the contradictions emerging from the "Social Question" that seemed to plague Spain after its imperial downfall. This romantic and autocratic Don Quixote-type personality of the water missionary became a literary hero in some of the period's regenerationist novels. César Moncado, the protagonist in the novel of failed revivalism (*César o nada*) by Pío Baroja written in 1909–1910, is an archtypical example of one such hydraulic, quasi Nietzschean missionary (Henn 2003). The hero, an agricultural and hydraulic regenerationist, is determined to create a municipal democracy in the Castilian village of *Castro Duro* by using his personal power to redistribute land and promote a plan for irrigation and reforestation (Baroja

[1909–1910] 1965). This requires, Baroja maintains through the words of his hero, "the destruction of *los caciques* (local traditional political elites), the termination of the power of the rich, the submission of the bourgeoisie . . . it will redistribute the land to the peasants . . . and *my dictatorship* will break the network of ownership and of theocracy"[29] (Baroja [1909–1910] 1965, 24). The figure of the strong, personal, and voluntarist, quasi-fascist leader as the pivotal actor in transforming the socio-environmental order is staged here as a key actor in breaking the control and power of the traditional elites. The novel narrates his desperate attempts to transform both land and traditional social-political configurations. He is elected to the *Cortes Generales*, or Cortes [Spanish parliament], constructs irrigation canals, plants trees, and builds schools. However, in the end, the hero fails in his aspirations to defeat the landed aristocracy and its allies. The novelist narrates, in a defeatist and cynical tone, the final victory of arid, conservative, and counter-revolutionary rural Spain: "Today, Castro Duro [the village of the hero] has now definitively abandoned its pretensions to live. . . . The springs have dried up, the school has closed, the trees . . . were pulled up. The people emigrate, but Castro Duro will continue living with its venerated traditions and it sacrosanct principles . . . sleeping under the sun, in the middle of its fields without irrigation"[30] (Baroja [1909–1910] 1965, 379).

In a comparable mode, the hydraulic missionary of Macías Picavea's novel *Tierra de Campos*, Manolo Bermejo, who the author at some point calls the *Christ of Valdécastro* (Macías Picavea 1896, 227) [the village of the hero], has to fight a similar and equally desperate struggle against tradition: "Castilian agriculture does not need the politics of the lousy advocates and the opportunists [referring to traditional elites], but it needs a hydraulic policy . . . irrigation . . . intensive agriculture"[31] (Macías Picavea 1896, 224).

After a brief period during which the hero tries to mobilize the village against the power, the protectionism, and conservatism of the large landowners of the drylands, the village ends up returning to its old habits and immobility. The hero is left behind—isolated, humiliated, and ruined—embracing "the dead soul of this poor fatherland" (Macías Picavea 1896, 325). These novels contemplate the fate of the land if the revivalist mission does not succeed, while already foreshadowing the formidable obstacles and the almost unavoidable failure of the project.[32] The desperation, the emphasis on the role of the enlightened male leader who pursues his mission doggedly and against all odds, and the stubborn resistance of traditional forces already hint at the later emergence of the Falangist ideology and fascist victory.

The failure of the hydraulic project during the early years of the twentieth century resulted exactly in what the regenerationists tried to prevent: sharpening crisis, culminating in civil war and the installation of a dictatorship. The latter, ironically, mirrors the qualities and characteristics of the heroes—the iron surgeons—that appear in the novels of Macías Picavea and Baroja. General Franco would indeed push through some of the reforms—and in particular internal colonization and the hydraulic remaking of Spain—the regenerationists had advocated (see chapters 5 and 6).

Purification and the Transformation of Nature: Hydraulic Engineers as Producers of Socio-Nature

The hydraulic intervention to create a waterscape supportive of the modernizing desires of the revivalists without questioning the social and political foundations of the existing class structure and social order was very much based on a respect for "natural" laws and conditions. The latter were assumed to be or thought of as intrinsically stable, balanced, equitable, and harmonious. The hydraulic engineering mission consisted primarily in "restoring" the "perturbed" equilibrium of the erratic hydrological cycles in Spain. Of course, this endeavor required a significant scientific and engineering enterprise, first in terms of understanding and analyzing nature's "laws," and second in using these insights to work toward a restoration of the "innate" harmonious development of nature. The moral, economic, and cultural "disorder" and "imbalances" of the country at the time paralleled the "disorder" in Spain's erratic hydraulic geography. Both needed to be restored and rebalanced, as nature's innate laws suggested, to produce a socially harmonious development. Two threads have to be woven together in this context: on the one hand the pivotal position of a particular group of scientists in the hydraulic arena, the Corps of Engineers (Villanueva Larraya 1991), and on the other hand the changing visions concerning the scientific management of the terrestrial part of the hydrological cycle. Both, in turn, were linked to the rising prominence of hydraulic issues on the socio-political agenda at the turn of the century.

The often-vitriolic dispute over the scientific management of the hydraulic nexus unfolded primarily between the civil engineers and the forestry specialists. For the former, large-scale hydro-infrastructure became the pivotal axis for efficient water management, while the latter considered centuries of deforestation and soil erosion in the mountainous regions of Spain as the main problem and advocated reforestation as the key strategy for improving the hydraulic balance and regenerating the land. The gloomy regenerationist imaginary of a barren and infertile Spain pointed to the "hatred of water and trees" as a key problem. A "national dendrophilia" therefore ought to replace the barren socio-physical landscape of Spain (Casado de Otaola 2010, 105). The reforestation of the mountainous regions that had been denuded for centuries as a result of the felling of trees for building the imperial fleet, for mining purposes, and for extending agricultural areas, became a priority in the eyes of the regenerationist forester. A hydraulic politics, so the argument went, could only succeed if combined with an intensive program of reforestation (Gómez Mendoza 1992b). The latter was seen as a vital ingredient of a battle against torrential run-off, increasing the water-capturing capacity of the land, stopping erosion, and facilitating the control of river flows (Casals Costa 1988, 1996). Joaquin Costa echoed this view in his impassionate defense of a "forested" regenerationism. For him, the key national question is the "reconquest of the soil through the tree" (Costa 1912, 13). In 1915, for example, the regenerationist writer, Julio Senador Gómez (1915, 115), repeated the plight of a deforested Spain: "And away from the basins of the rivers, everything is dead, burned, naked, crumpled, and dusty. It is the earth dying of thirst; it is because there are no trees; and as there are no trees, there is no water, no life, and everywhere you encounter horrors that looks as if they are the result of some geological cataclysm."[33]

The scientific management of forests, tirelessly pursued by the Ingenieros de Montes (the Corps of Forestry Engineers), served both economic and ecological-hydrological purposes. For many of them, a hydraulic politics that did not give center stage to the need for reforestation was anathema. Senador Gómez (1915, 119) reflected an opinion shared by many foresters when he claimed "what is needed is not a water policy, which means nothing, but a forestry policy, which means everything."[34]

The regenerationist foresters campaigned tirelessly for a reforestation policy as the backbone for regeneration and a necessary foundation for an effective hydraulic transformation of the country. A leading engineer and one of the founding fathers of Spanish Natural Park policy, the Catalonian Rafael Puig y Valls became one of the protagonists of reforestation who would not "correct the work of nature by replacing wooded areas with concrete constructions" (Puig y Valls 1898). He founded the *Sociedad Amigos de la Fiesta del Árbol* (Society of the Friends of the Tree Festival), which celebrated the first Arbor Day (Tree Festival) in Barcelona, an idea he had picked up during his travels in the United States (Puig y Valls 1909). The arcadian ideal of the foresters and the imaginary of reforested Spain as the nexus for rehabilitation, development, the repopulation of the countryside, and a more stable hydro-ecological configuration clashed with a hydraulic engineering vision that centered on large-scale infrastructure and a profound techno-natural restructuring the socio-physical matrix of Spain.

The animosity between hydraulic engineers and foresters would rage for decades, but eventually the "missionaries of concrete" would become the leading architects of Spain's socio-natural transformations. Despite a series of policy initiatives and sustained activism from the forestry engineers, the "love for trees" that galvanized parts of the regenerationist desires would be either incorporated into a National Park policy on the one hand or articulated with the hydraulic configuration on the other (Gómez Mendoza and Ortega Cantero 1987; Gómez Mendoza and Matal Olma 2002). Eventually, reforestation policies would become aligned with the hydraulic policy and an extension of river basin management (Serrada 2000) and integrated within the national hydrological planning framework (Gómez Mendoza 1989). Indeed, with the establishment of river basin management units (see chapter 4), forestry became an extension of, but ultimately subservient to, the hydraulic engineering mission whose ultimate objective was of course not just about improving water balances and protection against erosion and floods, but primarily aimed at providing the necessary hydraulic resources for agriculture, industry, and energy. Hydraulic policy did not just serve the purpose of restoring the nation's hydraulic balance and health but would become the cornerstone for the country's industrialization and urbanization too. Indeed, hydroelectrical production, satisfying the thirst of rapidly growing cities, and maintaining a thriving private construction sector would shape much of hydraulic intervention despite the continuing discursive attention to agriculture, reforestation, and internal colonization. The Corps of Engineers would ultimately dominate the making of the hydraulic landscape and become the master engineers of Spain's hydro-social transformation.

The Corps of Engineers, founded in 1799, was (and remains) the professional collective responsible for the development and implementation of public works. The great disaster of April 30, 1802, when the Dam of Puentes caved in and 608 people lost their lives, gave a great impetus to reenforcing engineering science and skills, and propelled the Corps to one of the leading professional associations in the country. In 1835, the engineering association was designated as *Cuerpo Especial* (Special Corps) within the state administration. It was a highly elitist, intellectualist, "high-cultured," male-dominated, socially homogeneous, and exclusive corporatist organization that took a leading role in Spanish politics and development over the centuries (Mateu Bellés 1995). Closely associated with the state and in charge of all the public works, the Corps of Engineers was not only the executive arm of the state's public work's program, but it also became a core actor in co-shaping state policies and interventions over the years. The decision-making structure was hierarchical and all key managerial and institutional bodies, such as the *Junta Consultiva de las Obras Públicas*, the hydrological divisions, the provincial headquarters, and ad hoc study commissions, were exclusively "manned" by engineers.

Mobilizing a technical-engineering discourse, the Corps of Engineers (and its key and influential publication *Revista de Obras Públicas*) would become a leading voice in support of large-scale hydraulic works and their vital relevance to regenerate the country. The Corps considered its members to be the key advocates and protagonists of the public initiatives for the hydraulic regeneration of the country and in designing its politics. It sought a significant degree of autonomy from state control and operated in the margins of all direct social action and conflict (Ramos Gorostiza 2001, 15).

At the turn of the century, the *ROP* would insist on the urgency of a hydraulic program, advocate major infrastructural investments, and widely publicize each initiative taken by the state. For example, Antonio Morales Amores would, in the aftermath of the 1898 debacle, systemically defend the regenerationist movement and advocate the preparation of a coherent national hydraulic policy. In 1899, he wrote in the *Revista de Obras Públicas*:

> The initiative and intervention of the State to realize numerous constructions of dams and irrigation canals is absolutely indispensable if we are to give to the agricultural and industrial wealth that regenerationist impulse which public opinion has demanded with so much insistence and persistence at these times. When, after having lost the territories that once were considered to be the base for the wealth of the Metropolis, we have to return our gaze to the land of the fathers in order to extract from it those many resources it offers, resources that have been abandoned during many centuries.[35] (Morales Amores 1899a, 164)

The planning and implementation of a system of dams and irrigation canals to foment the industrial and agricultural development of the country has to happen "now or never," the same engineer argued in the subsequent issue of the *ROP* (Morales Amores 1899b, 176). The state was called upon to take charge of the preparation and implementation of a national plan for dams and irrigation canals through which "the government would inaugurate an era of regeneration in this unfortunate land. ... It breaks the lethargy of the dormant life that we lived during so many centuries" (Morales Amores 1899a, 165). Regulating the irregular flow of the rivers, providing irrigation water and preventing water from being "lost in the immensity of the seas" were, for engineers too, vital preconditions for regenerating the country. The Corps of Engineers put its technical and scientific potential at the disposal of this national project and they would, systematically and enthusiastically, endorse and embrace each successive initiative by the state to take "the project" forward (Sáenz Ridruejo 1999, 10). The engineers considered their expertise to be particularly important to improve the regulation of rivers and to develop hydroelectrical systems to power Spain's modernization. The mobilization and propagation of a new technology that was seen as emblematically modern, namely reenforced concrete, would be vital to the success of this endeavor (Sáenz Ridruejo 2000). This was neither a moment to hesitate, nor to perpetuate the immobility that had characterized the nation's modernizing politics this far. A new plan and decisive action was required (Morales Amores 1899c).

In line with the emergent scientific discourse on orography and river basin structure and dynamics, the engineering community argued for the foundation of engineering and managerial intervention on the basis of the "natural" integrated water flow of a water basin, rather than on the basis of historically and socially formed administrative regions (see figure 3.1). The emergent geographical regionalization overlaid the traditional political-administrative divisions of the country, forcing a reordering of the territory on the basis of the country's orographic structure. The engineers, in turn, portrayed the latter as the crucial planning unit for hydraulic interventions. Cano García succinctly summarizes how this "scientific" perspective fused seamlessly with a political argument over the reorganization of the territorial-political structure of the country (Cano García 1992, 312): "To revert to the great orographic delimitation for organizing the division of the land represents a contribution made from within the strict field of our discipline [engineering] and at the same time, at least initially, it shows the abandoning of traditional political divisions and the importance of other perspectives and concepts."[36]

This scientific and natural division provided an apparently enduring and universal scale for territorial organization in lieu of the more recent political and historical scales associated with political-administrative boundaries. As T. C. Smith argues, "The identity of the drainage basin seemed to offer a concrete and 'natural' unit which could profitably replace political units as the aerial context for geographical study" (Smith 1969, 20). Jean Brunhes, whose geographical work was highly influential in Spain at the time, insisted on the water basin as the foundation for the organization

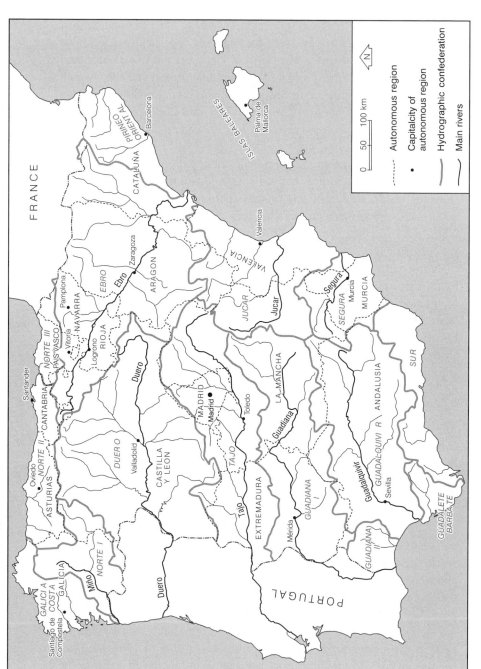

Figure 3.1

of the land since "water is the sovereign wealth of the state and its people" (Brunhes 1920, 93). Such a view was widely recounted in Spain, and its arguments were rallied in defense of a new orographic-administrative organization of the territory. The engineers, weary of what they considered the sterile political ideological feud between the two main parties who alternated, and thus basically shared, power, considered the river basin as "their" domain. The struggle to establish river basin authorities with significant political power was, for them, also a struggle to acquire power, to push through their agenda in light of formidable opposition by the traditional political elites who occupied the "traditional" territorial-political units (municipalities, provinces, the state). The debate and struggle over river basin authorities, fought through the mobilization of a scientific discourse, was also, and perhaps foremost, a struggle over scale, the production of a new territorial political scale, and the occupation of these new sites of power.

The history of the delimitation of hydrological divisions is indeed infused with the influence of the regenerationist discourse on the one hand and the scientific insights gained from hydrology and orography on the other. The attempt to "naturalize" political territorial organization was part and parcel of a strategy by the modernizers to challenge existing social and political power geometries. The construction of and command over a new territorial scale might permit them to implement their vision and bypass more traditional and reactionary power configurations. The complex history of the formation of river basin authorities and their articulation with other political forms of territorial organization, in particular the national state, is a long, complicated, and tortuous one. The river basins would become the scale par excellence through which the modernizers tried to undermine or erode the powers of the more traditional provincial or national state bodies. Therefore, the struggle over the territorial organization of intervention expresses the political power struggles between traditionalists and reformers. While the river basin defenders would become the founding "fathers" of the regenerationist agenda, the traditional elites held to the existing administrative territorial structure of power. The regenerationist engineers thereby incorporated the naturalized river basins into their political project. Capturing the scale of the river basin as the geographical basis for exercising control and power over the organization, planning, and reconstruction of the hydraulic edifice was one of the central arenas through which the power of the traditionalists (and the scales over which they exercised hegemonic control) was challenged. In fact, this rescaling of the state and the articulation of different scales of governance became one of the great arenas of struggle for control and power. For the engineering community and the modernizers, the scale of the river basin became the battleground over which political and social conflict was fought. The modernizers attempted to take hold of the hydrological divisions and tried to develop them as pivotal institutions for instigating the hydrological revolution, while the national scale remained more firmly in the hands of the traditional elites. The bumpy history of the hydrological divisions, which will be discussed further in chapter 4, records this struggle. The instability of their administrative and political organization reflects the relative power of the traditional power brokers. It would take until Franco's dictatorship before this issue was resolved when the hydrological dream and its intellectual bearers (the engineers) were aligned with and incorporated in a new organic-fascist structure of the state (Gómez Mendoza and Ortega Cantero 1992).

Conclusions

In sum, first, the regenerationist agenda(s) maintained that the restoration of wealth in Spain should be based on the knowledge of the laws and balances of nature. Second, this restoration required the correction of defects imposed by the geography of the country and particularly the "imbalances in its climatic and hydraulic regimes" (Gómez Mendoza and Ortega Cantero 1992, 173) and "the remaking of the geography of the fatherland." And third, only the central public authorities could carry out this enterprise of geographical rectification because of its range and importance.

The hydraulic mission was seen as the solution to the social and political problems facing Spain at the turn of the century. Hydraulic politics became the mediator to achieve key social and political objectives. The intensification of land use would increase productivity, thereby improving both the material conditions of the peasants and the competitive position of Spain in Europe. The increasing value of irrigated land, in turn, would improve fiscal revenue and remedy the structural problems of recurrent state fiscal deficits and debt. Moreover, the "natural" tendency of irrigated lands to fractionalize into smaller units would keep more people on the land (thereby slowing down the rural exodus), undermine the power and position of the *latifundistas*, and turn landless peasants gradually into smallholders. In this way, the social problem that Spain faced could be solved. The assumed political alliance between smallholders and some modernizing segments of the rising urban bourgeoisie would facilitate "the liquidation of the oligarchic *caciquismo* regime of Restoration Spain and announce the definitive consolidation of a real liberal-democratic state in Spain" (Mateu González 2002, 44). Failing this, social tensions were bound to intensify, and struggle, if not civil war, would be the likely outcome. Ironically, of course, the voluntarist, powerful, and autocratic hydraulic engineer pursuing a voluntarist program of imposed reform foreshadowed the fascist (Falangist) ideology. The latter would gain momentum from the early 1920s onward, first with the dictatorship of Primo de Rivera and later with Franco. The failure of hydraulic politics in the early decades of the twentieth century announced what Costa and his literary allies had feared and desperately tried to prevent. Although the debates at the turn of the century indicated a desire to regenerate Spain, conservative forces prevented its actual implementation and social tension intensified, further destabilizing an already highly fragmented and divisive society. The centralizing fascist regimes that emerged from this turmoil could finally push through the production of a new geography, a new nature, and a new waterscape, something the regenerationists of the turn of the century had so desperately advocated, but failed to accomplish.

Indeed, the regenerationist spirit and its movement was an arcadian fantasy (Harrison 2000b), a mythical vision without a people, without political roots. It aspired to be above politics, did not include the demands of workers or the peasants and did not strive to ally progressive forces. It reveled in a dream of land, progress, and a unified great Spain while discarding aspirations nurtured by many regionalists for a more devolved state. As Javier Varela argued, Joaquin Costa was "a failed prophet, a messiah without a people" (Varela 1997). Not surprisingly, therefore, clear anti-democratic schemes combined with charismatic leadership would emerge as part of the regenerationist agenda (Tuñón De Lara 1971, 75).

I have attempted in this chapter to reconstruct multiple and often-contradictory narratives that span a broad range of apparently separate instances such as engineering, politics, economics, culture, science, nature, ideology, and discourse. A new tapestry of sense and meaning is quilted together, one that imagines nature, land, and people in ways distinctly different from earlier imaginaries. Such radical imaginary, and the conflicting relations that search for its materialization, became a vital ingredient in the slow but tumultuous reordering of socio-physical space and the production of a new socio-environmental landscape; landscapes that are simultaneously physical and social, that reflect historical-geographical struggles and social power geometries, and that interiorize the flux and dynamics of socio-spatial change. Geographical conditions, this chapter argued, are reconstructed as the outcome of a process of production in which both nature and society are fused together in a way that render them inseparable, producing a restless "hybrid" quasi-object in which material, representational, and symbolic practices are welded together. Doing geography then implies the excavation and reconstruction of the contested process of the "production of nature." This excavation is precisely what I shall turn to in the next chapter.

4 Chronicle of a Death Foretold: The Failure of Early Twentieth-Century Hydraulic Modernization

De te fabula narratur[1]

The revivalism explored in chapter 3 did indeed galvanize most of twentieth-century political-economic thought and modernizing of hydro-social practices. The hydraulic desire of the regenerationists laid the foundations for reimagining Spain's development and modernization throughout the twentieth century, but it would take the best part of the century for the actual socio-environmental transformation designed by the high priests of the hydraulic dream to be realized. Indeed, the great transformation envisaged by the early modernizers would be smothered in the political and socioeconomic turmoil that characterized the first decades of the twentieth century. Much of this revolved directly around the controversies over the hydro-modernization project. Despite the recurrent discourse of hydraulic modernization, very little was actually achieved during the early years. Unstable governments, serious financial and economic problems in the aftermath of the colonial debacle, recurrent droughts and intensifying social conflict threw all manner of spanners in the works.

In this chapter, I shall briefly chronicle the key initiatives, the often-desperate attempts and "minor successes" that characterized the early trajectory of transforming Spain's hydro-social geographies. First, the nineteenth century's rather unsuccessful attempts to facilitate and foster private initiative will be discussed. In a context of spiraling crisis, attention turned increasingly to the need for a collective enterprise in which the state would take a lead role. Indeed, by the end of the century, hydraulic politics had acquired a mythical status capable of overhauling and reconstructing the nation's socio-geographical condition, but one that required the proactive role of the state. The torturous process of producing the territorial basis for comprehensive and integrated water politics, namely the establishment of river basin authorities, will be considered in light of the subtle and not so subtle power struggles that unfolded over the water nexus. The successive hydrological plans and policies to enhance irrigation—and their relatively small successes—will be considered.

Second, the post-1898 period unfolds as a chronicle of a death foretold. The regenerationists' worst fears—as well as some of their dreams and aspirations—would come true. Accelerated modernization, they argued, would be the only course of action to avoid further social disintegration and intensifying political-economic conflict. Yet, achieving the necessary transformations to embark on a sustained course of hydro-modernization required the concerted and authoritative action of an "iron surgeon," a benevolent and voluntarist leader of and for all the people. It required one who could contain the endless disputes and petty babbles of the two ruling parties and govern in the interest of the people, a leader who would stand above politics, displacing both parliament and quarrelling traditional political elites. As it turned out, the regenerationists got both.

When General Miguel Primo de Rivera abolished the Cortes and took the helm of the state in 1923, he set into motion a concerted effort to build a modern hydro-social infrastructure. However, it was already too little too late: ideological conflicts had intensified, social struggles were endemic, the fear of the elites for revolutionary change was too deep, and the desire of the impoverished workers and peasants for change was too real to halt. The maelstrom of modernity and the heterogeneous forces animating the political-ecological process had begun their descent into what would become the most bloody and intense civil war in Europe in the twentieth century.

Water as Fetish: The Unspeakable Truth of the Land

Space and land framed the fundamental problems that confronted Spain at the turn of the century. Spain was of course not unique in this respect. The scramble for Africa among budding European empires, the expansionist desires of some of the U.S. elites, the massive problems of large landholdings throughout the Mediterranean combined with impoverished masses of landless peasants, and the agitation of both anarchists and socialists around questions of land redistribution suggest the pivotal role land and space played at the time. In addition, the growing belief among Spain's elites that self-rule, sovereign development (*autarchia*), and self-sufficiency (*autarkeia*) went necessarily hand in hand as vital conditions for both independence and development (see Saraiva and Wise 2010) fueled the obsession with maximizing the productivity of land. In Spain, the problem of land, after the imperial debacle, became a thoroughly politicized internal affair, the axis around which much of the political debate unfolded. Yet it was rarely explicitly acknowledged as such, with the possible exception of the anarchist movement whose popularity grew rapidly both among landless peasants in the South and parts of the nascent industrial proletariat in the North. The debate over water and the need for the inauguration of a hydraulic politics became the stand-in, the terrain on which the disavowed hard kernel of the land problem was articulated. Radical land reform was the unspoken "truth," the censored theme in mainstream politics. Water politics became its mirror image, the domain on to which the social land problem was projected. It is exactly this disavowal of the centrality of land that underpinned the desire of hydraulic regenerationism and turned water into a viciously contested terrain, one that expressed the antagonistic positions of the traditional and aspiring progressive elites.

Indeed, hydraulic regenerationism, Alfonso Ortí argues, embodied an ideological mystification whereby hydrological plans and expectations were considered to offer a solution for the "agrarian question" and, consequently, for the social problems of Spain (Ortí 1984, 12). This agrarian question ultimately resided in the vexed problem of land ownership and access. Combined with the acute agricultural crisis of the turn of the century, the agrarian question spearheaded the concern of both modernizers and traditionalists. Successful industrialization and sustained rural to urban migration depended on "liberating" peasants from the land while securing a reliable and affordable urban food supply. Formulating the plight of the impoverished peasants and the modernizing needs of the country in terms of the "hydraulic desire" of arid Spain and the desolation of its landscape assembled all manner of unlikely allies to rally around the water nexus: H_2O was enrolled as the material and symbolic bearer of the fate of the land, one that offered a discursive formulation whereby pressing socioeconomic problems and necessary reforms could be posited in nonconflicting and nonantagonistic terms. The hydraulic fix—so it was argued—would intensify agriculture, increase productivity, permit the colonization of newly irrigated lands, facilitate minor land reform, and improve Spain's international competitive standing without having to consider the politically sensitive question of radical land reform. Water policy became the synecdoche and stand-in for dealing with the land and agrarian question that constituted the key nexus of Spain's lamentable socioeconomic condition. While being discussed in many circles, land reform became, as it were, a rapidly vanishing mediator to the extent that the hydraulic solution increasingly imposed itself as the one around which often-unlikely allies could assemble. The traditional political elites as well as the *latifundistas*, the large landowners, were obviously opposed to any type of land reform that threatened their political and economic interests. However, the radical left was equally skeptical about the merits of land reform (short of collectivization of the land). For communists and anarchists alike, land reform reeked, in its dominant formulation, too much of an attempt to instill a "bourgeois" logic in the peasantry by turning landless laborers into a land-owning smallholder class. For the left, such a process was seen as a pioneering form of capitalism that would hasten the transition from a largely feudal to a liberal-capitalist order. Moreover, empowering peasants by turning them into petty commodity producers was also seen as a threat by those, both on the left and the right of the political spectrum, who considered the state to be the pivotal arena for organizing and steering socioeconomic change. Neither the liberal reformers nor the conservatives or the Leftist revolutionaries took on the question of land reform directly.

Joaquin Costa's hydraulic regenerationism permitted displacing the vexed land issue and became "a collective myth in the service of a national politics" (Ortí 1984). Hydraulic politics as conceived by the regenerationists offered a way out of the contradictions described earlier by formulating the modernization problem in terms of availability, access, and distribution of water. This, so they argued, would solve simultaneously the land problem without threatening the position of large landholders, the agricultural crisis, ailing productivity, and the social problems of rural Spain. Their position was further strengthened as a result of a series of recurrent droughts that affected harvests directly and signaled that hunger and malnutrition were directly related to water availability rather than to the highly unequal distribution of land ownership and access. The drought of 1904–1905 was particularly severe and propelled the hydro-debate to new levels of intensity (Harrison 1973). In addition to focusing on political and social concerns about water, it put the state in a central position as the key articulator and implementer of the hydraulic dream. The process through which water was freed from the feudal shackles that tied it firmly to patrimonial control is what we shall turn to next.

Depatrimonialization of Water: The Failure of Private Initiative

In line with international sentiment, the nineteenth-century liberal revolution and the subsequent Restoration period had prioritized private initiative as the medium through which to modernize Spain's economy. In addition, the growing demand for water and the requirements for a more efficient and equitable distribution of irrigation waters necessitated a fundamental change in the legal status and appropriation rights of water. The liberal revolution in Spain (approximately 1811–1873), which had attempted an institutional (anti-)feudal restructuring to promote capitalist forms of ownership and the circulation of goods as commodities, extended also to what Maluquer de Motes called the "depatrimonialization of water" (Maluquer de Motes 1983). Indeed, as with land, water did not have the characteristics of a privately owned and tradable good, but rather of a personal or patrimonial good (Macías and Ojeda 1989). In line with Roman law, water was considered part of the "Public Domain," rendering access to water a right for all as a public good. Yet, in practice, solid private concessional rights regulated water access (Moreu 1999). The principles of eminent domain and the existence of a complex system of seigniorial rights over water prevented or blocked throughout much of the nineteenth century the development of productive activities through private initiative.

Throughout the nineteenth century, the Cortes took a series of legal initiatives that gradually changed the feudal water principles of eminent domain to a concessionary system whereby the state would regulate water access rights to those who wished to undertake irrigation or other hydraulic projects. Already in 1812, the Cortes passed a law that changed the status of water from eminent to public domain (Maluquer de Motes 1983). Most of the subsequent water laws (of 1819 and 1849) were primarily aimed at removing legal and fiscal obstacles for the private mobilization of water resources and at assuring legal certainty to the users/owners. Water appropriation and use was increasingly transferable and could be allocated according to productive use rather than distributed on the basis of seigniorial rights. While the state would organize the concessionary system, it was up to the private sector to propose and implement water projects; a view supported by a religious belief in the productive powers of private initiative. As Matteu Gonzáles attests, the state still reveled in a "geographical optimism" that combined a presumed abundance of water with an unshakable belief in the virtues of a liberal politics (Mateu González 2002, 40). All efforts were geared toward eliminating the institutional framework inherited from the *Ancien Régime* and at providing a legislative body that would transform private capital into the engine of irrigation-based development (Garrabou 1997, 142).

The series of incoherent and disjointed legal reforms of the first half of the nineteenth century were codified, integrated into, and amplified by the law of 1866[2]—an extraordinary legal achievement and at the time unique in Europe—that defined water as a public good and affirmed the role of the state in regulating a system of concessionary rights to permit the allocation of water to its optimal uses (Melgarejo Moreno 2000). This was followed by the first Water Law of 1879,[3] the key legal framework that declared natural water flows (basically surface water) to be a common good of the state. This law would regulate water, its access, and its distribution throughout the twentieth century until the passing of the new Water Law of 1985.

These legal initiatives were followed in subsequent years by a series of laws that regulated and tried to promote private initiative in the construction of public works, of dams, canals, and irrigation projects.[4] These initiatives codified and further entrenched the systems of water concessions along with a series of generous incentives (Gil Olcina 2001). The key features included assigning secure access rights to irrigating communities and to dam and canal development companies, and introduced a system of water payments (*un canon*) by those who received water for irrigation purposes. In cases of nonpayment, the farmers were obliged to sell the land to the irrigating organizations at the price of nonirrigated land (Martín Retortillo 1963). The liberal vision eschewed offering direct state subsidies, but the recognized reticence of the private sector to invest in projects with a long gestation period, uncertain long-term returns, and huge up-front capital costs prompted the state to devise all manner of indirect support and income streams for potential investors. Furthermore, returns on investment would only start kicking in a few years after the initiation of the project while still demanding huge initial capital outlays and yielding uncertain returns. Moreover, the advance of dam irrigation technology made the scale and scope of possible projects increasingly larger, further compounding the investment problem (Villanueva Larraya 1991, 20–27). During the presentation of new proposals for subsidizing the construction of dams and canals, the then Minister of Development José Luis Albareda lamented the failure of earlier attempts to kickstart the productivist transformation of the hydro-social landscape: "none of the companies that have tried to build irrigation canals in our country saw their efforts crowned with a successful completion." He identified a series of problems: "the high costs of the infrastructure because of the special condition of the rivers. . . . The scarcity of private capital and the absence of a spirit of association between capitalist and landowners . . . the low density of the population . . . the payments for irrigating the land . . . the traditional practices of the farmers" (cited in Villanueva Larraya 1991, 47). Of course, he forgot to mention in his list the deeply ingrained, almost religious, belief in the fertilizing forces of liberal ideas and free enterprise, continuous political instability and civil war, the deepening economic and agricultural crisis, the military costs of maintaining and managing the empire, the high cost of servicing an imperial elite bureaucracy, and the associated chronic budgetary problems of the state. In addition, many of the traditional forces, like large landowners, were positively opposed to large-scale hydraulic works, which they saw as a threat to their political and economic position.

Despite the radical change in legal frameworks and political efforts, very little of real significance was achieved during the second half of the nineteenth century. Half-hearted and quarreling political elites and a reluctant private sector failed to initiate the necessary transformation that would assure that "not a single drop of water would reach the sea." Very few hydraulic works of major significance were constructed during that period. Quenching the thirst of large cities and dealing with rapidly deteriorating sanitary conditions in the industrializing urban environments became, as elsewhere in Europe at the time, a key priority. The water provision of Madrid was inaugurated in 1858 (Canal de Isabel II) (March Corbella 2010), Barcelona had begun to upgrade its water supply (Guàrdia, Rosselló, and Garriga 2014), and the Canal de Urgel was completed in 1862. With 102 kilometers of canals and 3,000 kilometers of irrigation ditches, another 70,000 hectares of land had been irrigated. All other projects together added another 300 kilometers of canals and approximately 70,000 hectares to the total expanse of irrigated lands. By the end of the century about 1.2 million hectares of Spanish agriculture was irrigated, 70 percent permanently. This constituted about 4.5 percent of cultivated land, but only 2.4 percent of potentially cultivable land (Villanueva Larraya 1991, 79–88). Many concessions were awarded, numerous projects launched; most either did not even get started or collapsed at some stage during the implementation process. All in all, these projects fell short of the great expectations vested in the liberal dream of transforming the productive capacity of the land through private initiative. Moreover, the condition of state finances was disastrous. The state's indebtedness was gigantic. Interest payments on the national debt gobbled up 27 percent of state spending. If one adds the gigantic cost of maintaining the military presence and war efforts in the colonies, all manner of payments to political elites, the police, and the bureaucracy, and the maintenance of the clergy, two-thirds of state spending was already accounted for (Tortella 1981, 140). During the period from 1860 to 1900, only about 9.5 percent of the budget was allocated to the Ministry of Development, most of which went to roads and railroad investment and maintenance, with only an average of 3.9 percent of the total development budget dedicated to hydraulic projects (Villanueva Larraya 1991, 88–89).

By the end of the century and after two decades of intensifying social disintegration, political turmoil, military defeat, and economic crisis, the tide began to change and the call for a more radical transformation of Spain became louder. The regenerationists, who had begun to agitate for a more concerted national and public involvement, rallied around the need for a real hydraulic politics in which the state would take center stage. However, there was a major snag, for it was the oligarchic and quasi-feudal elites who occupied hegemonically the state apparatus.

The State as Master Socio-Environmental Engineer

To irrigate is to govern.
—J. Costa, cited in F. J. González Martín, "Filosofía del Derecho y Regeracionismo Político en el Concepto de Estado de Joquín Costa"

Since the early 1890s, and spearheaded by the Agricultural Chamber of Aragón[5] of which Costa was an active member, a greater number of social and political forces allied around the view that the ambition to regenerate Spain through a geographical project necessitated concerted action and collective control. The regenerationists welcomed the liberation of international markets and the demise of nineteenth-century protectionism under which the dryland *latifundistas* of Spain (mainly South and Central) flourished. By 1880, trade liberalization had plunged agriculture into a deep crisis in the aftermath of the expansion of the United States' wheat export boom. The traditional landed bourgeoisie was economically weakened as a result, but its members' political commitment to maintain their power on both national and local scales did not abate. This control permitted the continuation, if not reinforcement, of a strong protectionist economic policy framework.

Nevertheless, central state intervention to produce a nature amenable to the requirements of a modernized, competitive, and irrigated agriculture was considered essential (Ortega 1975). The state should intervene directly in the implementation of hydraulic works that would permit it to "remake the geography of the fatherland" (Costa [1892] 1975, cited in del Moral Ituarte 1998, 121), revive the national economy, and "regenerate the people (*la raza*)" (del Moral Ituarte 1998, 121). These hydraulic politics were for Costa a way to insert Spain within a European socio-spatial framework after its loss of influence in the Americas on the basis of a rural development vision that combined a Rousseauian ideal with a small-scaled, independent, and democratic peasant society. The promotion of the rural ideal on the basis of a petty bourgeois ideology would become the spinal cord of the liberal state and the route to the Europeanization of the nation (Nadal Reimat 1981, 139; Fernández Clemente 1990).

The regenerationist vision, then, faced with the failure of privatized and commodified water to operate as an efficient allocation and productive instrument promoted the emergence of a collective spirit ("illusion" in the words of Alfonso Ortí 1984, 14). The latter implied that the supply of the necessary quantities of water was only possible under a public and socialized form or, in other words, through the state. This collective and state-led vision would eventually also include the state-led production of "great hydraulic works" (Ortega 1975; Villanueva Larraya 1991). In sum, the regenerationists turned to the state—after the failure of the liberal project to defeat the feudal elites—as the agent that could generate a sufficiently large volume of capital to mobilize Spain's natural resources. Moreover, for Costa, the productivist modernization by means of the hydraulic motor would in fact consolidate the liberal state in Spain. In short, a free-market based, intensive, and productive national economy, whose growth would accelerate on a par with other northern European states, necessitated a transformation in the state in a double and deeply contradictory sense. First, power relations within the state apparatus needed to change in favor of a more modernist alliance of smallholders, industrialists, and modernizing engineers and, second, the state needed to support the social reform that such a revolution required so that the grand hydraulic works could lay the foundation for a modernizing Spain. Surely, this contradiction turned out to be irreconcilable. These two tasks were of course mutually dependent. Strong traditional forces fought to maintain control over key state functions and prevented the rise to political power of the nascent petty-owners and middle classes. This firm hold of the traditional conservative elites blocked most attempts to modernize the social economy. The mosaic of contradictory forces and the resistance of the traditionalists would stall state-led modernizing efforts, resulting in more acute and openly fought social antagonism throughout the first two decades of the century. These would eventually pave the way for dictatorial regimes from the 1920s onward.

Despite the collectivist discourse of much of the regenerationist literature, it remained deeply committed to a project of entering an international capitalist market. Although the state needed to take central control over water and forests, it was to do so on the basis of a land ownership structure that was essentially private and market-led (see also Fernández Clemente 1990). The hydraulic agenda of Costa and his colleagues was spiced with revolutionary zeal and ardor but ultimately advocated a reformist route for development against the stronghold of an anti-reformist, and economically and culturally conservative and protectionist elite. In sum, two models or imaginaries of capitalist accumulation, with evidently different supporting social groups and allies, crystallized around the hydraulic debates at the time. The social, political, and ecological consequences and implications of these two models would differ fundamentally even while sharing an organic vision of the world. On the one hand, the traditionalists defended a protectionist economic stance and the continuation of existing political and social power relations. On the other, the regenerationists advocated a more liberal perspective, a rapid modernization of the economy, and a transformation of socio-political power relations. It is a struggle choreographed by two different class positions: traditional elites associated with the land-empire nexus and an emerging liberal middle class. The issue of land ownership and the role of water therein revolved around the question of who would own and control what part of the land and its waters. For the regenerationists, for whom small-scale family-based ownership constituted the way ahead, the hydraulic route was an essential precondition, while the limited possibilities for accumulation pointed to the state as the only body that could generate the required investment funds on the one hand, and push through the necessary reforms in the face of strong and sustained opposition from the landed aristocracy on the other (Ortega 1975; Ortega Cantero 1992). At the same time, the very support of at least some sections of the old elites could be secured via this reformist route, since it did not threaten their fundamental rights as landowners and it defended rural power against the rising tide of the urban industrial elites and proletariat. This was indeed central to forging the support of the dominant Catholic groups that defended a solidaristic and organic model of social cohesion.

Enrolling the State in Hydraulic Metabolism: Plans, Plans, Plans

The regenerationist imaginary quilted a discursive chain of signifiers around water as the pivot that captured Spain's condition. A radically new imaginary was stitched together about Spain's plight, its causes, and the necessary path of action. This new imaginary placed the state in pole position. The question now was how to make this imaginary constitution infuse the state apparatus and its practices.[6] In this section and the one following, we show how the radical imaginary of this hydraulic project and its social mission was enrolled within the state, thereby transforming the very functioning and configuration of the state itself. This proved to be a thorny process, one that was riddled with difficulties and problems along the way. In the first decades of the twentieth century—until the Civil War—the discursive apparatus of the regenerationists and its key signifying chains entered the vocabulary of many of the leading state actors, some of whom would desperately try to turn image into reality. However, while the imaginary of the regenerationists was gradually adopted, political practices and policies remained anchored around the traditional Restoration power geometries.

The opening salvos for transforming the state's position and role in the development of hydraulic works and with it, the attempt to make the state take the lead role in the country's development, were fired in April 1899, less than a year after *El Desastre*. Drought was wrecking the country's agricultural production, and this situation would only worsen over the subsequent years. The fire came from two directions simultaneously. On the one hand, the Corps of Engineers explicitly endorsed and advocated the regenerationist agenda and many would spend their professional engineering lives in advancing its cause (Sáenz Ridruejo 2000). On the other, Rafael Gasset, owner of the most influential independent newspaper, *El Imparcial*, launched in the pages of his newspaper a passionate discussion of and plea for embracing state-led hydraulic works as the way forward to get Spain out of the doldrums it was in. Even in the Cortes, the delegates agreed that hydraulic public works were an utmost necessity and, despite great internal debate and disagreement, instructed the Government of Spain to present a national plan for hydraulic works within one year.

The Corps of Engineers had already advocated the need to have state-based dam and irrigation projects to counter the agricultural crisis and to reverse the dismal results of private water works initiatives in a doctrinal article published in 1896 (Cardenal 1896). In the meantime, the engineers had begun to work on the preparation of a national plan for hydraulic works. A few years later, on April 7, 1899, the General Commission of the Corps of Engineers presented this first "plan" for public hydraulic works, prepared by hydraulic engineers on the initiative of the Corps. This *Avance de un Plan General de Pantanos y Canales de Riego* (Proposal for a General Plan of Dams and Irrigation Canals) was presented to and discussed with the minister of development (Cuerpo de Ingenieros 1899a). On the eve of the twentieth century, the *Avance* (the Proposal) became the emblematic first coherent proposal for the planning and development of the nation's water resources on the basis of large-scale and state-initiated hydraulic projects. Symbolically, the plan marked the beginning of what would later be termed "hydraulic structuralism." Although the plan did not offer an integrated vision and was based on limited data, it presented an inventory of projects to be implemented, proposing a total of 170 dams and 65 canals that, combined, would lead to the extension of irrigated land by more than 1.5 million hectares. The cost to the state of the full implementation of the project was estimated at around 400 million pesetas. The plan charted the contours of what, over the next century, would shape discussion, determine the material form, and outline the institutional and organizational contours of national water planning and development.

In the subsequent weeks, months, and years, the *Revista de Obras Públicas*, the official periodical and mouthpiece of the Corps of Engineers, published a series of passionate and hard-hitting articles, supporting their plan and pressuring the state to take action. Their arguments were highly influential among the technocratic modernizers within the state apparatus. These contributions lamented the fate of Spain, diagnosed the problem in terms of the insufficient use of water in raising the productivity of the land, and the failure of the state to take on this task. The influential engineering society put its considerable shoulders squarely under the regenerationist project and became one of its great protagonists. The lead article of the issue of April 13, 1899, for example, solemnly declared the Corps' intention to contribute to "the national labor of economic reconstitution" (Cuerpo de Ingenieros 1899e, 131). By mobilizing the expertise of local engineers and putting together the plan—and in symbolic reference to the disaster of the U.S.-Spanish war—they emphatically stated that "we [The Corps] have begun the war of peace, the war of labor, the fight for progress that, instead of devastating, restores; of destroying, builds; of draining, enriches"[7] (Cuerpo de Ingenieros 1899e, 131). The analogy of water, both a source of (steam) energy for industrial development and for enhancing agricultural production, was invoked to symbolize the need to shift the gaze and the vital energies of Spain from pursuing the colonial adventure to the transformation of the fatherland:

During several centuries, Spain sought to acquire more space. And in the way water needs to become vapor to do so, and in the process consume energy, so as space became more extended, more energy leaked from the metropolis. Then, the coldness of the thanklessness has turned to condense the steam, and when concentrating the molecules again in their original container, will give back again those lost energies that today each Spaniard will use in the transformation of his mother country.

After the deep convulsions suffered, our country directs its anxious gaze towards every element of life that can restore those lost energies, and today places its hopes in irrigation. But, however floating these aspirations are, it was needed to channel them, to focus the mind, to orient all the forces and to apply them to the same objective at a given time in order to convert them into a powerful national drive.[8] (Cuerpo de Ingenieros 1899e, 132)

The subsequent week, the *Revista de Obras Públicas* repeated its clarion call, once again reinforcing its position and trying to rally, in a muscled tone, the technocratic and political elites of the country: "the conquest of new territories has been realized more by the efficiency of the roads, canals and ports than by the power of arms" (Cuerpo de Ingenieros 1899d, 147). The territorial conquest was now one aimed at developing public works and promoting and using the newly available engineering skills and technologies—like reinforced concrete and new construction methods (see Becerril Bustamante 2008)—as the basis for rapid modernization and, through that, securing advantageous competitive positions in industry and commerce.

The pressure was sustained all along those tumultuous months of April, May, and June 1899. In the issue of April 27, for example, Antonio Morales Amores kept drumming to the same beat:

> Our most renowned fertile valleys would be susceptible of a more intensive culture if they could take advantage in the summer of at least part of the waters that in the other seasons of the year rush away before the eyes of the farmer to be lost in the immensity of the seas, if not causing damages or even great disasters. . . . It is the State's task then to take care of the study and implementation of a plan of dams and irrigation canals; and for the Government to open up an epoch of regeneration in this wretched land.[9] (Morales Amores 1899a, 165)

To reiterate the urgency of the situation, the same author concluded his contribution in the subsequent issue as follows: "The dilemma is clear: Now or Never" (Morales Amores 1899b, 176); one week later, attention shifts to the great dangers awaiting the country if urgent and immediate state action is further postponed (Morales Amores 1899c). In addition, the argument in favor of public hydraulic works is extended to include the importance of hydroelectricity for the development of industry and the need to transport over long distances those forces that nature does not offer where needed. On May 25, 1899, the *Revista* published a passionate call urging the political leaders to embark on a hydraulic politics, defending it as a unifying patriotic mission and, once again, pledging the full political, technological, and intellectual effort of the Corps of Engineers to the realization of the project (Cuerpo de Ingenieros 1899c).

This vocal and public endorsement of a national hydraulic interventionism coincided with a sustained effort in parts of the press to raise public consciousness of the issues and pressure the government to take action. Rafael Gasset, once a conservative party candidate to the Cortes and owner-editor of the leading independent newspaper *El Imparcial,* started a series of influential articles on April 7, 1899, under the telling headline of *Para la Nueva Política* (In Defense of the New Politics).[10] Over the years, *El Imparcial* had already offered its pages—and in particular its widely read cultural section—to contributions from regenerationist intellectuals like Unamuno, Maeztu, Azorín, and Baroja. In the articles, Gasset launched a great propaganda effort and, as he called it, a public awareness crusade to popularize the importance of regenerating the country through large-scale state-led hydraulic public works (Sánchez Illán 1898). The vision was a great one, the work to be done enormous, the challenges considerable, he said, but there was no other way to deal with the defects of the Spanish soil. As Gasset put it: "The work to dam and withhold the water of our rivers that with torrential and devastating velocity bridges the distance between the springs and the sea is gigantic, and as such, is faced with overwhelming obstacles . . . but if only small issues were undertaken, simple and plain, we would never remedy the rectifiable defects of our soil. We would never raise our feeble and weak production"[11] (Gasset, cited in Cuerpo de Ingenieros 1899e: 132).

For Gasset, this hydraulic politics would be a "patriotic labor," "transcendental," and deal with "the aridity of the Spanish soil" (Villanueva Larraya 1987, 443). Endorsing the Proposal prepared by the engineers, the editor of *El Imparcial* urged his colleagues and readers to support this "because only a supreme effort of the now weak public opinion is capable of overcoming the resistances and obstacles that always oppose all that is great" (Gasset, cited in Villanueva Larraya 1987, 444). Fully aware of the formidable political opposition to radical change, he pleaded for the government to instruct its engineers to develop a national plan and implement the necessary projects, and insisted that "all political influences" should be avoided in the development of the plan. Both he and the engineers shared the view that an exquisitely and rationally argued but purely technocratic plan would be sufficiently persuasive to forge an alliance between often hostile and unlike political and social actors strong enough to rally around this national project.

Indeed, only a few weeks later, on May 11, 1899, the main political leaders across parties seemed to agree with this. In the council of ministers presided over by Queen-Regent Maria Christina, Prime Minister Francisco Silvela argued, "Drought is a national disease that acquires the proportions of an Egyptian plague and brings today the death of our miserable agricultural wealth" and insisted on Spain's need to remedy with all means possible the effects of the drought, and execute the vast plan for dams and irrigation canals (Cuerpo de Ingenieros 1899b, 191). On July 11, 1899, a number of deputies[12] presented a statement to the Congress of Deputies, once again insisting that the state should take on board the construction of dams and canals and pleading that a general plan of such works should be presented to the Cortes within a year (Villanueva Larraya 1991, 108).

Yet again, the obstacles and difficulties were expounded on at length. First, the hegemonic position and postures of the advocates of a liberal market economy were condemned, insisting that even the heartland of market liberalism, the United Kingdom (often referred to as *Manchesterismo*) had abandoned those liberal principles for the construction of hydraulic works. Second, all manner of technical problems hindered the plan's implementation, but they could be solved by mobilizing the engineers' capacities and skills. Finally, and perhaps most important, the objections and hostility of the landowners to irrigation were foregrounded as a major obstacle, something that could only be tackled, so they argued, if the state would offer irrigation water free of charge for the immediate use of the landowners.[13]

This flood of petitions, proposals, statements, demands, and declarations turned hydraulic regenerationism into the talk of the day around the turn of the century. The concerted efforts to spread the message were of course aimed at rallying larger segments of the public around its cause. The political elites had no other choice than to include some of these visions in their programs and engage in what Tuñón de Lara dubbed their "pseudo-regenerationist" politics (Tuñón de Lara 1974, 73–79). Francisco Silvela, who assumed the office of prime minister in March 1899, defined his administration as a "Government of National Regeneration." However, the extraordinary state debt after the military defeat (in 1898, more than 40 percent of state spending went to finance the colonial war [Solé Villalonga 1967]), the demands of the traditional elites to finance the rearmament of the decimated military infrastructure, the desperate attempts of the government to reduce the gigantic deficit, and the unwillingness of the traditionalist to concede even minor reforms all mitigated against the swift implementation of the envisaged hydraulic policy.

To show his government's commitment to the regenerationist agenda, Silvela appointed Rafael Gasset to the post of minister of agriculture in April 1900, together with another regenerationist, Pablo de Alzola, who accepted the post of director general of public works. Gasset would be minister in nine governments between 1900 and 1923 (of the more than twenty that succeeded each other in staccato tempo during that period), switching in the process from the Conservative to the Liberal Party (in 1905) (Villanueva Larraya 1990). In his indefatigable manner, he set to the task of initiating a state hydraulic politics. Just one month after his appointment, on May 11, 1900, a Royal Decree was signed. The decree reorganized the national Hydrological Service and created seven Hydrological Divisions and a central inspectorate (*Jefe del Servicio*) that were instructed to study, propose, and develop a plan of hydraulic public works. This general plan was to be ready by December 31, 1900. In an optimistic and highly emotive tone, many welcomed the decree as the "real beginning of the regeneration of Spain" (Cardenal 1901). The text of the decree extolled the virtues of water works in redressing the fate of the country. For the first time, the uneven distribution of water was directly linked to questions of justice and the sense of despair that characterized so much of the literature of the time: "We do not deny, of course, the geographical facts that history brought us. The Sahara sends us devastating winds via the coast of the Levant, blows of fever from the African continent that render everything dry and sterile. Rainwater, although not rare, is distributed with such injustice and irregularity"[14] (Gasset 1899, II).

The *Plan de Obras Hidráulicas* (POH) (Plan for Hydraulic Works [PHW]) was finally published with the Royal Decree of April 25, 1902 and published in full in the journal of the Corps of Engineers (NN 1902a,b,c). The PHW would be the emblematic moment that announced the inauguration of Spain's "Hydraulic Politics." It would frame, shape, and guide the transformation of the country's hydro-social environment for the new century (Ortega Cantero 1992). A long list of 205 projects with a grand total of 296 planned dam and canals was presented, all exclusively oriented toward the improvement and extension of irrigation. About 1.2 million hectares of dryland were to be transformed into irrigated land and another 300,000 hectares of already irrigated land were to be improved, making the total surface of irrigated agriculture close to 2.5 million hectares. The total cost at the time was estimated to be 412 million pesetas (Gil Olcina 2010). Table 4.1 summarizes the projects envisaged by the plan.

The approved plan was ultimately not much more than a long list of individual projects, often highly localized, without a clear integrative vision, lacking in data, disjointed, and, above all, without clear technical, managerial, budgetary, and economic implementation guidelines. Later, in 1933, Lorenzo Pardo, key author of the First National Water Plan (discussion follows) and legendary figure in Spain's water debates, was highly critical in his assessment of the plan of 1902: "the country hoped for a plan of Hydraulic Politics, and what resulted was a catalogue of dams and irrigation canals, isolated projects without any relationship to the wider river basin" (Lorenzo Pardo [1933] 1999, 24–25). The projects signaled a weak understanding of geographical and hydrographical realities. Moreover, there was a complete absence of regulating dams for managing the entire flow of the river and of considerations to foster hydroelectrical production (Lorenzo Pardo [1933] 1999, 38–41). Nevertheless, the plan was a milestone in Spain's hydraulic history, particularly because it enshrined and reinforced

Table 4.1
The plan of hydraulic works of 1902

Hydrological division	Irrigated surface	Projects	Planned canals and dams
Eastern Pyrenees	30,000	7	18
Ebro	327,000	30	51
Júcar	143,470	15	18
Segura	14,300	9	12
Southern Spain	9,400	6	7
Guadalquivir	177,900	20	38
Guadiana	406,602	43	43
Tajo	181,850	22	47
Duero	152,600	43	51
Miño and Cantábria	25,900	10	11
Total	1,469,022	205	296

Sources: Villanueva Larraya 1991, 121; NN 1902a,b,c.

the vision that the state should take a central role in the planning and construction of water works. However, irrespective of its intrinsic qualities or otherwise, very few of the plan's proposed projects were realized in the subsequent decades. Indeed, the state remained fairly passive. Although no one opposed the program explicitly, the reality was that "the governments showed a constant and devastating passive resistance" (Villanueva Larraya 1987, 450). Yet, the power of the plan to hegemonize and generalize the regenerationist vision should not be underestimated.

The subsequent governments all endorsed, at least rhetorically, the hydraulic agenda. In 1905, the Liberal Party invited Rafael Gasset to be minister of public works (which he accepted) and embraced the hydraulic politics he stood for, undoubtedly prompted by the devastating drought of 1904–1905 that exacerbated an already acute agricultural crisis. However, despite a flurry of legal initiatives and official declarations, progress was excruciatingly slow for reasons already discussed. A despairing Gasset demanded "an energetic action that would end this politics of words, which keeps us today as poor and weak as the day that the Paris Treaty was signed"[15] (cited in Villanueva Larraya 1987, 451). In light of the slow progress, a new, less ambitious plan was approved in 1909, which basically comprised of a subset of the originally planned projects in the 1902 plan, but included both a budget and a time horizon for implementation of eight years (NN 1909). In addition, Gasset, as the minister of development in the liberal government of 1911, prepared a new law to facilitate and encourage state-led hydraulic works. The law signaled the end of the controversy over whether the state or private initiative should take the lead in the implementation of the hydraulic politics. From then onward, state intervention would depend on the political will to commit the necessary public resources to mobilizing Spain's water for irrigation purposes (Villanueva Larraya 1991, 161). Interestingly enough, the law focused on large-scale irrigation works (in contrast to small-scale, locally initiated and peasant-led irrigation projects) and was conspicuously silent about the contentious issue of colonization in those areas that were transformed from drylands to irrigated fields, once again skillfully meandering around the politically thorny issues of land reform.

The engineering community agitated tirelessly in support of Gasset's relentless efforts to kickstart his beloved regenerationist plans in the face of formidable opposition from a wide range of actors. The law of 1911 was published in full in their journal, insisting again that the production of a new Spain through a hydrological transformation was the only option open. Examples from the U.K., India, Egypt, Mexico, and the United States were invoked to signal the success of state-led public works (NN 1911). Insisting on the scientific evidence for implementing such policy and in increasingly desperate terms, the Corps of Engineers lamented the political stalemate and the continuous obstacles put in the way of real modernization as a result of continuous party in-fighting and the cult of *"personalismo,"* an exaggerated emphasis on great man initiatives at the expense of a collective and coordinated effort (González Quijano 1913). In 1916, in the midst of World War I, a revised Plan for Hydraulic Works, this time coupled with an extraordinary budget, was presented to the Cortes. It covered a total of 156 projects and aimed at expanding irrigation by 571,441 hectares, just a fraction of what was envisaged in the first plan of 1902. The war—in which Spain was neutral—was invoked as an ideal moment for Spain to unify and catch up as the other warring nations were in the process of destroying the material and social basis of their own modernization (Gasset 1916). In 1919, yet another plan was passed, yet another variation on the same theme, showing the increasing desperation of the hydraulic politics' defenders at the slow progress.

By the early 1920s, hydraulic politics had been ingrained in the imagination of many Spanish leaders. The relentless campaigning, the endless flow of reports, plans, speeches, analyses, and proposals had produced a new imaginary for water interwoven with metonymic signifiers for modernization, development, regeneration, irrigation, engineering technologies, steel and concrete, dams, integration, social cohesion, and national pride. This imaginary constitution, however, was not paralleled by a similar enthusiasm for the realization of the dream. The Ministry of Development's budget accounted for about 9.4 percent of state spending during the first decade of the twentieth century, and increased to about 13.5 percent in 1919 and an exceptional 21.4 percent in 1920. On average 73 percent of this budget was earmarked for public works, although hydraulic works only accounted for 6–9 percent of this (Villanueva Larraya 1991, 210–211). In terms of actual developments, the picture was even gloomier: by 1926, only a miserly eighteen of the projects envisaged in the plan of 1902 had been realized (Lorenzo Pardo [1933] 1999) and no more than 150,000 hectares of new land was irrigated. In the meantime, social agitation from both left- and right-wing forces intensified, often leading to violent uprising and bloody repression. Likewise regionalist desires and demands for autonomy grew, political infighting and instability continued relentlessly, military efforts to maintain at least some vestige of imperial grandeur by holding desperately to a recalcitrant Spanish Morocco drained the budget, and fiscal reform was systematically blocked. While Gasset, much of the Corps of Engineers, and other modernizing groups put all their efforts in pushing through their techno-natural dreams for a prosperous Spain, the social and political edifice around them was crumbling.

On September 13, 1923, General Miguel Primo de Rivera took power, suspended the Cortes, and established the first dictatorship of the twentieth century in Spain. During the dictatorship, the enlightened technocrats and the dictatorial regime would begin to build the institutional configurations and assemblage of forces through which—in the second half of the twentieth century—Spain's hydro-social edifice would be produced.

The Struggle over Scale: The Hydrological Divisions

The failure of early twentieth-century modernization resided principally in the solid control that traditional political forces and economic elites exercised over all territorial scales of government: municipalities, provinces, and the state. For decades, the modernizers desperately tried to carve out an administrative-territorial basis that would permit a certain degree of autonomy and control to push through their techno-modernizing ideals. The struggle over the making of such a territorial power base in the form of river basin authorities exemplifies the intense political power struggles between traditionalists and modernizers. The negotiation of scale and the science versus politics debate over the scaling of hydraulic intervention and planning raged for almost a century before the current structure was put into place (Cano García 1992; Mateu Bellés 1995). All along, the engineers and their allies defended the river basin as the appropriate scale for the techno-administrative management of water on the basis of "scientific" arguments whereby the unity of the basin and the integrated nature of its water flows were mobilized to legitimize the appropriateness of their political inauguration. With the dictatorship of Primo de Rivera, the first autonomous *confederaciones sindicales hidrográficas* (CSH) (hydrographic confederations) were established, many decades after the hydraulic engineers and regenerationists had begun to militate in their favor.

The choreography of the struggle to establish river basin authorities testifies to the intense political power struggles and the territorial nexus around which this was fought. The Water Act of 1879 had established that all surface water was common property, managed by the state. This also implied the need to create administrative structures to perform these managerial tasks (Giasanta 1999). The first hydrological divisions were established by Royal Decree on July 29, 1865 and, from the very beginning, they were considered to be major instruments for the economic modernization process (Albaredo Page 1871). Miguel de Cervantes, for example, lauded their establishment as instruments to study Spain's water and mobilize its potential: "Great, immense will be the increase in our country's wealth the day that all the water it has will be supplied to those places where it can be used best." In order to achieve this, he continued, "Great enterprises have to be in charge of the gigantic construction of dams and canals . . . and the landowners need to implement the necessary work to irrigate their lands" (de Cervantes López 1865, 263). Some of these divisions more or less coincided with major river basins (Ebro, Tajo, Duero). Others (like in the South) had a much closer correspondence to provincial boundaries. All were named after the provincial capital city where the head-office was located (Mateu Bellés 1994). Their role was to serve as an institutional basis for collecting statistical data for the study and research of the water cycle. These surveys could then be used as inputs to the real power holders (provincial head offices for public works, special ad hoc commissions, or private initiatives) (del Moral Ituarte 1995). In many ways, these hydrological divisions were very early and innovative attempts to organize water policies on the basis of integrated river basin management, something that would only become more or less generalized worldwide in the late twentieth century.

The hydrological divisions that had been established by the law of 1865 were abolished in 1871 in the midst of the turmoil of the liberal-republican revolutionary period, a "lamentable suppression" that "will be remembered as a fatal one in the history of Public Works," the Corps of Engineers claimed (de Palau Catalá 1875, 234). "There are now several engineers . . . rendered idle by Government decree, and there are also many rivers that . . . gush to the sea or are hidden in the sands,"[16] the article concluded (de Palau Catalá 1875, 238). A year later, the *Revista* demanded the immediate reinstatement of the hydrological divisions. They alone, the engineers argued, could guarantee a proper geographical and statistical analysis of water supplies and needs. It was the engineers who could in an impartial and authoritative manner propose adequate, "truthful," and "beneficial" plans, judge projects, and offer the "seal of confidence" conferred by "the judicious and dispassionate assessment" of such public administration (García Hernández 1876, 172).

Budgetary problems were the officially stated reason for abolishing the hydrological divisions. More to the point, of course, was another reason: the traditional elites within the state apparatus were wary of parallel hydraulic administrations with significant power and of the considerable potential for raising the stature of the engineering professionals who would have been in control of the divisions' operations. After closure, the divisions' personnel and documents were transferred to the provincial offices of the Ministry of Development. The divisions were partly reestablished in 1876 during the Restoration, but their number was reduced to five (Duero, Tajo, Ebro, Guadiana, and Guadalquivir), then increased to seven (with the addition of Segura/Júcar and Miño) in 1881. They were again reduced to three in 1886 and finally abolished again in 1899 (Mateu Bellés 2004). Needless to say, this extraordinary instability and continuous reordering is symptomatic of the political struggles and conflicts that raged over the desirability of and the need for such new forms of territorial governance.

In 1899, the Hydrological Service was created, an organization within the state and with engineers working in each of the provinces, and replaced, in 1900, by seven divisions of hydraulics works, organized around the river basin as structuring principle, and overseen by a central service. This would also include forestry and reforestation plans and solidify the hydro-forestry nexus under the unifying canopy of the hydraulic frame. These divisions would be the administrative units to undertake the necessary studies for and oversee the implementation of the 1902 Plan Gasset. The Corps of Engineers cautiously welcomed this, "exactly at the time when the entire country believes that this could be one of the most adequate measures to support national wealth in a practical and efficient manner, offsetting the painful losses that Spain has recently suffered"[17] (NN 1899, 485). The Corps also insisted that only its continued effort could prevent this new service once again descending into "sterile" labor and enable "reason" (of course, *their* reason) to triumph (Cardenal 1901). The divisions were basically statistical and technical organizations and political power remained firmly in the hands of the political administrations, which supervised and executed the hydraulic works and financed and controlled the infrastructure programs. In addition, the torturous early formation of the divisions showed the limits and weakness of the successive governments of Restoration Spain (Mateu Bellés 1994; Sáenz Ridruejo 2001, 49). For the engineers and modernizers, however, quasi-autonomous and powerful divisions were an absolute necessity. The success or otherwise of hydraulic politics would stand or fall with the establishment of such territorial configurations, which would replace, at least in part, the existing lineage of traditional territorial political power configurations (NN 1904).

With the dictatorship of General Primo de Rivera, the "iron surgeon" who could confront the stronghold of the traditional parties and local *caciques*, the first regenerationist regime was established. During the dictatorship, powerful river basin authorities were finally put in place. It is indeed only from 1926 onward that the current *Confederaciones Sindicales Hidrográficas* were gradually established as quasi-autonomous organizations in charge of managing water as stipulated by the Water Act of 1879 (Giasanta 1999). The last of these ten *Confederaciones* was not established until 1961 (see figure 3.1). The consolidation of authoritarian state power and the inauguration of integrated river basin management unfolded hand in glove as the means through which hydraulic politics would shape the country's modernization. Powerful techno-managerial water institutions were established at the scale of the river basin, while the national scale both supported and nurtured these water management and hydraulic works organizations. As Miguel Beltrán put it, both the Primo de Rivera and Franco regimes wished to "empower an Administration of Engineers for the realization of State led public works. . . . The Government pretended to be nothing more than a committee of high-level functionaries who decided in consultation with the great Corps of Engineers what had to be done" (Beltrán Villalva 1996, 569). In their desire to substitute the political for the professional management of problems framed through a techno-managerial lens, the Corps approved and supported this trajectory to consolidate and increase its privileged position within the state (Ramos Gorostiza 2001, 15).

Too Little Too Late: The Regenerationist Dictatorship of Manuel Primo de Rivera and the Emergence of the Hydro-State, 1923–1930

What had proven impossible to achieve during the first decades of the twentieth century was firmly set in motion during its first dictatorship, although it would prove too little too late to save Spain from gliding further into the abyss that would lead to the Civil War. On September 13, 1923, General Manuel Primo de Rivera staged a successful coup d'état from his base in Catalonia in the midst of major agitation from leftist labor unions, dissolving the Cortes and then establishing, with the full approval of King Alfonso XIII, a Military Directory. The constitution of 1876—as well as civil liberties—was suspended. Many observers and protagonists, even some on the left of the political spectrum, welcomed the dictatorship as a means to unblock the inertia and deadlock of the political system. The Corps of Engineers cautiously welcomed the dictatorship of Primo de Rivera in 1923 (ROP 1923). In an editorial to mark the end of the dictatorship in 1929, they celebrated the grand works undertaken by the regime and the accomplishments of the dictatorship (ROP 1930, 2003a). The support of many of the modernizers for the new regime is captured well by the great and influential philosopher José Ortega y Gasset, writing in *El Sol* on November 27, 1923: "The alpha and omega of the task that the Military Directory has imposed is to make an end to the old politics. The purpose is so excellent, that there is no room for objections. The old politics must be ended"[18] (Ortega y Gasset 1983, 26). Schlomo Ben-Ami, eminent scholar of the dictatorship, argues that Primo de Rivera was presented as the legitimate heir of the regenerationist myth, the one who comes to cure the sick, removes the politicians and political parties from power, and begins the work of healing the country (Ben-Ami 1983, 58). His dictatorship was a syncretic one, eclectically combining a military tradition with Costa's mythical "iron surgeon," the need for a revolution from below and the urgency to defeat the militant left anarcho-syndicalists (Ben-Ami 1983, 84–86). The colonial disaster figured prominently in the ideological edifice. Only after a major effort of national reconstruction could "the pulse of Spain" be restored to inspire again Spain's international aspirations (Tamames and Casals 2004). José Antonio Primo de Rivera, the dictator's son, would later found and lead the Falange Española, the fascist party that would become one of the leading supporters of Franco. He became, after his execution in 1936, a powerful symbol of martyrdom for the fascist regime.

The regenerationist credentials of the authoritarian regime were secured by appointing a hydraulic engineer with impeccable conservative pedigree, Rafael Benjumea Burín, to the post of minister of development in December 1925, after the Directory of military commanders that ran the country was partly replaced by a Civilian Directory. King Alfonso XII had conferred the title of *Conde* (Earl) of Guadalhorce upon him in 1921 in honor of his engineering work in the Guadalhorce river basin and, in particular, for the completion of the hydroelectrical and irrigation complex around the Dam of El Chorro in Southern Spain (Martín Gaite 1983). As a shrewd investor, he had used the new state subsidies for hydraulic works to set up major hydroelectrical and irrigation companies and become a leading businessman (Boelens and Post Uiterweer 2013). Inspired by the example of Mussolini in Italy,[19] Benjumea would oversee the major program of public works that underpinned the economic vision of the new regime.[20] After the harsh initial years of the dictatorship, a program of infrastructure investment was initiated to kickstart the economy again. However, the massive loans required to finance these project, combined with a worsening international economic situation, resulted in massive inflation. The poor felt the consequences most acutely. By the end of the 1920s, the situation had deteriorated to such an extent that both the army and the king as well as many of the elites had lost confidence in the dictator's ability to pacify the country. He stepped down in 1930, announcing a period of great turmoil that would build up to the dramatic events of 1936.

The Conde del Guadalhorce called immediately on another engineer, Manuel Lorenzo Pardo, to develop proposals for the integrated development of river basins. Pardo, a key and enigmatic figure in the dramatic unfolding of Spain's hydro-social edifice and later author of the First National Plan of Hydraulic Works, had been working, since 1906, as a hydraulic engineer for the Division of Hydraulic Works of the Ebro River Basin. He had accumulated an extraordinary in-depth and intimate knowledge of the water economy of the Ebro River and had been centrally involved in some of the largest hydraulic projects at the time (see Lorenzo Pardo 1916, 1930). His vision extended the objectives and aspirations of the hydraulic community by insisting on the need to shift upward from a focus on individual projects to the integrated management of entire river basins. Such a bold framework, which he developed in detail for the Ebro River, would become the characteristic labor of the era and set a pathbreaking example and model for the management of Spain's national water system (Lorenzo Pardo 1931, 145). He argued for "the immediate conquest of a new Ebro" (Marcuello 1990, 82), a fully engineered and managed cyborg river.

The torrential and intermittent flow of the Ebro rendered only small parts of the river navigable, while the provision of water for both hydroelectrical generation and irrigation was limited and irregular. Only considering the hydro-dynamics of the entire basin could rectify this, Lorenzo Pardo argued. Such a vision required the introduction of what he termed "hyper-regulation" by means of a series of "hyper-dams," supported by a "scientific" view of the basin as a single unit, but managed with the participation of all the users in the context of a general planning framework. Such hyper-regulation would secure a regular flow of usable waters throughout the year. This comprehensive and integrated management required a new institution to which the national administration had to devolve full financial, technical, and managerial responsibility. Lorenzo Pardo would, in subsequent years, become a tireless "agitator" to defend and implement this notion. For example, in 1922, he pleaded with the then minister of development that "all hydraulic services should be centralized . . . in a single unit for each river basin; it is the only way in which this very important source of wealth will respond efficiently to the national interest"[21] (Lorenzo Pardo 1930, 225–226). The dictatorship would constitute the great moment of its implementation.

After he was appointed, one of Conde del Guadalhorce's first acts of governing was to call Lorenzo Pardo to Madrid for a meeting and instruct him to prepare urgently a concrete plan of action based on his vision of integrated river basin management (Marcuello 1990). Just a few weeks later, his proposals were discussed by the council of ministers. On March 5, 1926, the Royal Decree to establish *Confederaciones Sindicales Hidrográficas* (CSH) was signed by the King. The first of the *confederaciones* (river basin authority), for the Ebro River, was established the same day (Marcuello 1990, 155). Later that year the CSH for the Segura and for the Tajo basins were established, and in 1927 the Duero and the Guadalquivir authorities followed suit. The remaining ones would be established later. This "Copernican Revolution" in water politics pioneered, for the first time in the world, the integrated and participatory management of the full hydro-social cycle of the river basin (Fanlo Loras 1996, 105). The CSH mission was to organize the legal, concessionary, and administrative regulation of water in the basin and to take charge of the implementation of large hydraulic infrastructures, either by the state or with the support of municipalities, provinces, water users, irrigation communities, and other interested parties. The CSH operated in a semi-autonomous matter—at least in the early years—on the basis of four founding principles: the unity of the river basin as the ideal scale for the management of water resources, the water basin as integrated planning unit, the participation of water users in the management of the river basin, and the decentralization of state functions to the scale of the river basin. While the broader framework was centrally decided, water users (hydroelectrical companies and irrigators in particular) were welcomed to present their interests provided they did so in technical-managerial terms. In doing so, the autonomy of the CSH rested on a self-defined technocratic, apolitical position aimed solely at improving the uses of water for interested parties.

These CSH—the name alone suggests a decentralized organization of political-administrative authority—became the precursors of the great, leading examples of integrated resource development. Seven years later, on May 18, 1933, the Roosevelt administration established the Tennessee Valley Authority (TVA) in a concerted effort to mitigate the consequences of the Great Depression and inaugurate integrated economic development planning; the Spanish *Confederaciones* had already pioneered this form of integrated river basin development schemes. The parallels between the Ebro Confederation and the TVA are indeed striking (see Velarde Fuertes 1973; Fernández Clemente 1986). The CSH would also inspire comparable initiatives around the developing world (like in Mexico in the 1960s and 1970s) (Marcuello 1990, 175–178; Ekbladh 2002).

Manuel Lorenzo Pardo would, as expected, become the technical director of the Ebro Confederation. He immediately set to work to implement the principles he and other engineers had defended for so long. Between 1926 and the end of the Primo de Rivera dictatorship, the Ebro Confederation would become a shining example of effective and efficient integrated water management. Finally, the enlightened modernizers had conquered a territorial base from where they could begin to turn their regenerationist geographical imaginaries into material reality. It was also from the founding of the Ebro Confederation onward that the confederations were named after the river basins for which they were responsible. In addition, they acquired a certain political status with participation from the state, banks, the Cámaras de Comercio (Chambers of Commerce), and provincial authorities. At each stage the engineers took the lead roles. They became the activists of the regenerationist project through their legitimization of scientific knowledge and insights and their privileged position within the state apparatus (Mateu Bellés 1994, 1995). Territorial governance arrangements should, so the engineers argued, follow the scientific logic of "natural" boundaries rather than political-administrative ones.

However, constructing and commanding a territorial scale for the implementation of the hydraulic dream for Spain proved too little too late. The establishment of the CSH by a dictatorial regime had taken water management away from the idiosyncrasies of everyday political infighting and petty interests and transferred it to a managerial unit operating as a participatory techno-managerial outfit. Between 1926 and 1930, almost 200,000 hectares of new or improved irrigated land had been added in the Ebro river basin alone. However, it was not enough to turn the dire socioeconomic conditions around sufficiently to appease intensifying social conflicts, counteract the opposition of the traditional elites, and fight off the accusation of being an instrument of the dictatorship. The cards were indeed stacked unfavorably.

A State in the State and the Emergence of Enlightened Hydraulic Despotism

José Ramón Marcuello, Lorenzo Pardo's biographer, argued: "The energetic seed of the [Ebro] Confederation had hidden in its folds the germ, the lethal virus of its own destruction"[22] (Marcuello 1990, 193). The large landowners remained radically opposed to state intervention in the irrigation of their lands and, needless to say, the danger of expropriation. The powerful lobby that tirelessly advocated intense hydroelectrical development saw its influence curtailed. A more vocal political opposition to a waning dictatorial regime began to attack what was considered to be one of its great achievements. The most sustained critique centered on the arguments that the confederations operated as a "state within the state." The opposition forces lamented the confederations' financial and executive autonomy and their ability to set priorities through internal deliberative mechanisms, while the perception of pilfering, omnipotent, and unaccountable bureaucrats pursuing their own interests contributed to a general sense that the confederations were institutions beyond the law, the state, and the political and economic establishment. Published after the fall of the dictatorship, an editorial of the *Revista de la Confederación Sindical Hidrografica del Ebro* stated: "The Confederation has neither never been nor can be the exclusive patrimony of one party, not even of a political regime. Ever since its legal inauguration, it is an Institution that was born and lives to attend to the national needs and to promote and improve public wealth, a common goal of all ideologies and all efforts of the government"[23] (NN 1931).

The disavowal of the intensely political nature of the CSH and the attempts by Pardo and his allies to portray them as apolitical organizations working solely in the common interest placed them squarely into the center of multiple attacks from both the left and the right. The attacks culminated in all manner of accusations and innuendo including widespread rumors that the Ebro Confederation had bought diamonds for the wives of the engineers or that engineers had bought land and farms at undervalued prices. Moreover, the economic crisis of the late 1920s weakened the regime of Primo de Rivera. By the end of the 1930s, he stepped down in the midst of growing political turmoil. The King went into exile and the Second Republic was inaugurated.

As the proponents of the regenerationist project had begun to glimpse the contours of what an enlightened hydraulic despotism could achieve with the installation of its own territorial power base, they were not overly enthusiastic in welcoming the Republic. On May 1, 1931, the *Revista de Obras Públicas* published a formal and not excessively roaring endorsement of the newly established democratic regime. The Popular Front government of 1936 "was welcomed with even less enthusiasm" (Sáenz Ridruejo 2003, 11). All opposition forces considered the CSH as the symbol of the dictatorship. During the subsequent years and despite frequent changes of government (the Second Republic had no fewer than twenty-six governments between 1931 and 1938, oscillating between left- and right-wing coalitions), the CSH were radically transformed and aligned much more closely to the interest of the national state and its power choreography. Yet, the basic techno-managerial scale for implementing hydraulic interventions, the river basin authority, remained in place.

Aligning the national state would require another political revolution, this time one that would last until the mid-1970s. With the fall of the regenerationist dictator, the one who preferred "men of action" rather than "men of thought," Spain plunged into deep social and political turmoil, a process that had been brewing for several decades now and one that the regenerationists had so desperately tried to prevent. A modern, democratic, and civilized state, articulated around an Arcadian dream of fertile and irrigated *huertas* cultivated by prosperous and happy farmers had failed to materialize. It was a death that Costa and his allies had already warned of at the end of the nineteenth century.

Marching to Civil War while Reimagining the Nation's Waters

By the end of the Primo de Rivera dictatorship, river basin authorities had been established as quasi-autonomous techno-administrative bodies for the planning and implementation of the nation's hydraulic policy. However, the entrenched opposition to the new territorial power base around river basins signaled a profound distrust of the authorities from a range of different and often mutually conflicting political (regionalists, nationalists, republicans, monarchists, socialists, anarchists, etc.) or socioeconomic (large landowners, labor movement, peasants, etc.) interests. The political and economic instability after the dictatorship and the rapid succession of governments with different political allegiances shaped a kaleidoscopic tapestry of ever-changing positions, interests, and coalitions. Yet, throughout this tumultuous period that would culminate in the horror of the Civil War (1936–1939), the scalar geometry of hydraulic management was reordered such that the national scale and the scale of the river basin were aligned strategically. This, in turn, would set the frame that the dictatorship of Franco would mobilize effectively. In this process, water was enrolled in new ways. The imaginary of entangling Spain's waters was scaled up from a focus on the river basin to one in which water was imagined as a national hydro-social problem and configuration, while the rhetoric of water scarcity was increasingly replaced by an imaginary of national imbalance and hydro-social injustice.

Immediately after the inauguration of the Second Republic in 1931, the minister of development Alavaro de Albornoz (from the Radical Socialist Party) drastically restructured the river basin authorities. They are renamed as *Mancomunidades* (Fernández Clemente 2000). The preamble to the law of reform insisted indeed on the tutelage of and control by the central administration. The participatory assembly was abolished and, instead, a governing body appointed directly by the minister of development ran the affairs of the new authorities. A few months later, Lorenzo Pardo was replaced as director of the newly created *Mancomunidad Hidrográfica del Ebro* and moved to work under the national general director of public works.[24] With the first constitutional government of the Republic in 1931, the socialist minister Indalecio Prieto initiated a national planning system for hydraulic works, operating within and under the direction of the Department of Public Works. With this, Prieto consolidated the river basin authorities as a technical-administrative appendage and official service of the central Ministry of Public Works. In 1934, the name of the authorities was changed once again, this time to the one they still carry today, *Confederaciónes Hidrográficas*. Part of their original mandate to engage in land reform and colonization was removed. The Popular Front government of 1936, which offered far-reaching autonomy to Catalonia, would transfer all services and activities related to water and its infrastructure to the *Generalidad de Cataluña* (the Regional Government of Catalonia). During the brief Popular Front government, Catalonia was given full autonomy over "its" waters and associated hydraulic works. The centralizing forces of the Franco dictatorship would of course reverse the latter. In fact, many of the nationalists, both from the left and the right of the political spectrum, considered the original quasi-autonomous river basin authorities as a wedge mobilized by regionalists to gain greater independence for the regions.

Indalecio Prieto's initiatives would take up again where the Conde del Guadelhorce had left it in the early 1920s. In 1932, he appointed Lorenzo Pardo as head of the Section of Hydraulic Planning and instructed him to create an institute for the study and planning of water that would permit the design of a hydraulic politics for the nation as a whole. On February 22, 1933, the national Centre of Hydrographical Studies (*Centro de Estudios Hidrográficos* [CEH]) was inaugurated and its first mission was to prepare a National Plan of Hydraulic Works that would integrate and coordinate water planning and its implementation at the scale of the national territory. On May 31, 1933, Prieto signed the plan. It had been completed in a minimum of time (Lorenzo Pardo [1933] 1999).[25] This first national water plan, which the Cortes actually never officially approved, would not only set the parameters for water works and planning for the next sixty years, it also radically reimagined the hydro-social cycle. The plan staged and mobilized H_2O in a national framework, something that would be embraced fully by the later Franco regime. As Diaz Marta noted, this National Plan of Hydraulic Works was excellent in its "technical-scientific" assessment of the country's water economy and offered a realistic set of projections based on an integrated evaluation of national needs (Díaz-Marta Pinilla [1969] 1997, 54). Both the socialist minister Prieto and his successor from the *Partido Republicano Radical*, Rafael Guerra del Rio, enthusiastically endorsed the plan. For Prieto, it provided the foundation for a new Spain (Prieto 1933, 3). For Guerra del Rio, the plan offered "the valorization of the national patrimony and the rehabilitation of our maltreated economy . . . and inspires the regeneracionist labor of the Republic" (Guerra del Río 1933, 3–6). The plan offered an apolitical vision, he insisted, that stood above class and other interests, and around which all political parties could rally in a peaceful and democratic manner. Once again, water and its mobilization were hailed as vehicles around which all manner of interests could peacefully assemble. Of course, this would remain wishful thinking.

In line with the aspiration of the early regenerationists, the plan envisaged the construction of 215 new dams and canals, extending irrigation with an additional 1.3 million hectares, thereby doubling the surface of irrigated lands. This, in turn, would permit the repopulation of deserted areas and improve the self-sufficiency of Spanish food production while strengthening its international competitive position. The state would take full responsibility for the financing and implementation of these works. However, the plan's most important and significant contribution was its perspective on the integrated management of river basins in the context of a national assessment of regional water availability and requirements.

The hydraulic regulation of the entire basin would permit the integrated and coordinated planning of water for agriculture, energy, and other uses, and assure continuous supply, even in dry years. This would transform the imaginary of water from something to be considered, managed, and engineered within the unity of the basin to a national vision. The latter would be enshrined in what Pardo defined as the "Hydrological Disequilibrium" between the Atlantic and Mediterranean basins, a problem that could be "rectified" by inter-basin water transfers. Water transfers from the Ebro and, in particular, the Tajo River, to the basins of the Levant would solve the problem of systematic underprovision of water in the basins of the Southeast, a volume that Pardo estimated to be around 5,000 hm^3 (cubic hectometers) per year. It is precisely this number that will be repeated well into the twenty-first century as the "deficit" of the Mediterranean basins. The waters of the Segura basin, Pardo argued, were practically exhausted. Therefore, "the realization of the immense productive possibilities of the basin required an additional inflow of water from elsewhere. This posed the need to find new external resources to solve definitively the problem of the southeastern region of the peninsula"[26] (Grindley Moreno and Hernández Gómez-Arboleya 2010, 8). Minister Prieto stated in 1933 how a water transfer from the Tajo to the Segura basin constituted a "redeeming labor" for the nation (Lorenzo Pardo [1993] 1988). A national hydraulic imaginary was now envisaged, one linking national development with a national vision of how to organize the terrestrial flow of water. In the words of Pardo on the occasion of a speech he gave in Alicante in 1933:

> The phantasm of water transfer . . . is blown away in the face of the powerful blast of a supreme national usefulness. If one day we are masters of Spain's waters, which fall in sufficient quantity, but are unevenly delivered, and we can distribute them according to that usefulness, we will have accomplished the greatest act of sovereignty. The fine and melancholic spirit of the North, tempered by Castilian austerity, will be able to fertilize here in the Levant with the combination of the waters from over there and the sun here; and then, yes, national products will be yours that are capable to be taken to the to the last corner of the world, from America to the Far East, an embodiment of the Spanish soul.[27] (Lorenzo Pardo [1933] 1988, 68–89)

Where the early hydraulic regenerationists enrolled water in a discourse of scarcity, one at once produced by the climatic conditions that limited absolute availability of water in certain river basins and by the inefficient or limited engineering and use of the available water, Pardo's vision reimagined Spain's water economy by shifting the perspective from the river basin to a national territorial gaze. Seen from that scale, Spain's water problem became one of unequal distribution of abundant waters, able to fertilize the whole of the country. Rectifying this unevenness, soon to be dressed as rectifying socio-spatial inequalities and injustices, would become the legitimizing leitmotiv of hydraulic interventions.

Of course, the turmoil of the 1930s and 1940s would prevent the early realization of Pardo's national hydraulic dream. The fantasy, however, would galvanize much of the imaginaries around which General Franco and his allies would cement a hegemonic vision of autarchic national development and spatial integration. In his Plan of Public Works of 1940, Alfonso Peña Boeuf, minister of public works in the first post–Civil War fascist government, explicitly referred to the plan of 1933: "Its fundamental and foundational work will be very useful, and we believe that this must continue and be updated, and amended if required"[28] (cited in Urbistondo Echeverría 1984, 192). I shall now turn to explore how water is reimagined and rescripted in a particular fashion as part of the Francoist project. The early dream of the regenerationist would finally find its zenith in the materialization of Franco's liquid dream for Spain.

5 Paco El Rana's Wet Dream for Spain

On my shoulders rest all the problems of Spain: to forge the unity between the men and the lands of Spain, to bring water to the thirsty lands, . . . social justice; in short, the responsibility to make a new Spain.
—General Francisco Franco, extract from speech, April 29, 1961, Granada

To the so-called "generation of 98"—thinkers and "dilettantes"—an oppositional generation of men of action has arisen since 1935, whose realizations have resulted in the economic development of Spain.[1]
—Franco, interview in *Le Figaro*, June 13, 1958

By the 1930s, the conviction that Spain's modernization—after the colonial debacle—was predicated upon its internal socio-environmental transformation in which the mobilization and enrolling of H_2O would take a central place had been firmly and consensually ingrained in the elite's consciousness. As documented in chapter 4, the institutional arrangements as well as the will from some elites to push through such projects were put partially in place despite continuous opposition, political turmoil, and persistent economic difficulties. Yet, relatively little had been achieved in terms of actually transforming the hydro-techno-natural landscape. During the subsequent fascist period, Spain would undergo a veritable techno-natural revolution that would overhaul fundamentally the form, flow, and structure of the terrestrial hydro-social cycle. In this chapter and in chapter 6 I seek to document and substantiate further the notion of the production of socio-natures by excavating the processes of how Spain's modernization after the Civil War unfolded as a deeply and very specific scalar-geographical project. I shall focus on the momentous transformation of the hydraulic environment during the Franco period (1939–1975) and seek to reformulate Spain's socio-hydraulic reconstruction in the context of contradictory multi-scalar politics. Two theoretically interrelated arguments guide this endeavor. On the one hand, Franco's ideological-political mission was predicated upon national territorial integration, the eradication of regionalist or autonomist aspirations (Carr 1995), and a concerted discursive and physical process of cultural and material nationalist homogenization and modernization. A heterogeneous assemblage of social actors, political forces, and economic interests as well as a particular scripting and mobilization of water would amalgamate in a more or less coherent coalition of interests through which a new hydro-nature would be constructed. This chapter will focus on the construction of such national network of interests.

However, the implementation of Franco's hydro-modernizing program was paralleled by a geographical rescaling of the networks of interest on which Franco's power rested from a national visionary to a geoeconomic and geopolitical integration in the U.S.-led Western Alliance that emerged during the Cold War politics of the second half of the twentieth century. Chapter 6 will detail this scalar transformation. The hydroscape fused together a national project within reconfigured transnational geopolitical networks and their associated power choreography that had partitioned the world into separate political camps fighting for hegemony.

Under General Francisco Bahamonde Franco, more than six hundred dams, small and large, were constructed in Spain (Vallarino 1992, 67) (see figure 1.2). This led to a complete reengineering of the ten continental river basins of mainland Spain and the use of the Southern river basins "to the last drop of water," so that not a single drop would reach the ocean without having flowed through a sociotechnically engineered hydro-social process (like dams, pipes, pumping stations, irrigation networks, hydroelectric power generation, industrial and urban water networks, and the like). By the end of the dictatorship, the waters of the southern river basins were fully used. As Gomez De Pablos puts it, "During the two decennia after the Plan of 1940, the Spanish rivers were really created, with the construction or initiation of the principal regulatory works, on which a nationally integrated water politics was based"[2] (Gomez De Pablos 1973a, 242). Relatively speaking, Spain has the highest number of dams in the world (twenty-nine per million inhabitants), followed by the United States with twenty-three per million inhabitants (Llamas Madurga 1984). Under Franco's rule, Spain's hydraulic development reached its apogee (Pérez Picazo and Lemeunier 2000) and its logic continued long after the transition to democracy in 1978. Indeed, by 1993, the initial proposals for the new hydrological plan envisaged the most mammoth transformation of the Spanish waterscape to date by including a national water grid that would interconnect every mainland river basin and transfer a huge volume of water from the Northern basins to the Southern ones, where the local river basins had been "used to the last drop" and where the economy was facing considerable water "deficits" (see chapter 7). Franco's original vision of a nationally integrated and interconnected national orographic system would finally be completed.

The main focus of this chapter is on mapping the constellation of network of interests through which the Francoist project was organized and sustained, and examining the mobilization of the material forces of water within this assemblage. Indeed, the incoherent and conflicting social and political forces that had characterized the instability of the Republican governments were replaced by a more or less stable coalition of diverse interests around a national, if not nationalistic, project. The varied acting of water and its particular enrollment in a national development project was an integral part of this process of networked assembling. Despite internal conflict and differences, a range of diverse actors, practices, and discourses assembled to permit the rapid development of an unparalleled, revolutionized hydro-natural landscape, one that was to provide a pivotal axis for Spain's rapid modernization from the late 1950s onward. Moreover, a solidarist ideology, a drive toward modernization and the chosen (and partly imposed) autarchic economic policy during the early years of Franco's regime demanded an accelerated and intensified use of available domestic resources. The emphasis was on achieving national "redemption" and "restoration," while the social question—important during the first quarter of the century—moved to the background. By the time Franco died in 1975, virtually all river basins had been exploited to the full. Before we launch into the dynamics that permitted the material realization of Franco's wet dream for Spain, we shall first consider how the reimagining of the socioecological configuration of Spain's waters set the frame for its subsequent enactment during the dictatorship.

Enrolling H$_2$O: Nature's Injustices

Throughout the Franco years, water infrastructures and the transformation of the techno-natural edifice of Spain would be mobilized with relentless zeal by the propaganda machinery to such an extent that the popular nickname for General Franco, still ingrained in people's mind today, is *Paco el Rana* (Frankie the Frog). The most popular, omnipresent image of Franco during this period was of him being "on water" while inaugurating yet another hydro-technical project (see figure 5.1a, b). A passionate angler

Figure 5.1
Paco el Rana: the iconography of Franco's hydro-modernization. (a) "El Caudillo inaugurates the Dam of San Bartolomé" (June 3, 1942); (b) Generalissimo Franco contemplates a model of a hydraulic infrastructure (July 18, 1963).
Sources: (a) Ministerio de Educación, Cultura y Deporte, España, Archivo General de la Administración, Fondo "Medios de Comunicación Social del Estado" (MCSE), signatura 33-03305-00009-001. (b) Ministerio de Educación, Cultura y Deporte, España, Archivo General de la Administración, Fondo MCSE, signatura 33-03304-00010-001.

and ardent supporter of hydraulic works, "Frankie the Frog" consequently became a standard iconic gesture to ridicule Franco's "obsession" with water.

General Francisco Franco resolutely embraced part of the regenerationist discourse of the early twentieth century that centered on a strong state as the protagonist for the implementation of hydraulic modernization. The fascist national project would seamlessly weld together the national engineering dream of "balancing" Spain's uneven hydrology with a discourse and vision of social justice, and organic integrated national development. Water availability and distribution became articulated and experienced as a problem of state "voluntarism" rather than resulting from "natural" scarcity. If problems of scarcity existed, this was simply because of the state's incapacity to perform its functions adequately. State management of water generated a sense of unlimited potential availability. What was understood in earlier decades as a "natural" water limit became reinterpreted and scientifically defined as "deficit," "imbalance," or "disequilibrium" between the regionally desired volumes and the nationally available quantities. The dominant view of an uneven "natural" distribution of rainfall and water availability was increasingly resymbolized as a disequilibrium that required "rectification" (Sánchez de Toca 1911, 299–300): "This pluvial regime that is uncertain, exaggerated, excessively abundant during some months, and poor or zero during others" can, if well distributed, "sustain an intensely productive vegetation over the whole national territory" (del Prado y Palacio 1917, 67–68). An internal war against drought had to be fought so that "idle" rivers would provide "drink to the dry lands of Spain" by means of a "chirurgical remedy" to rebalance the socio-ecological matrix of the nation (Rodríguez Ferrero 2001, 126). Existing water scarcities were the mere result of inadequate state intervention to assure a better distribution of the nationally available waters. Of course, imagining abundant "national" waters but regional "imbalances" required a scalar reconfiguration that shifted the gaze from considering the hydraulic balances at the scale of the river basin to the national territorial scale. The latter, in turn, was predicated upon disavowing regional, if not regionalist, demands and regional hydraulic autonomy in favor of national water solidarity, whereby an integrated space could be rearticulated by transferring water from "surplus" to "deficit" river basins. It was the state, therefore, that had to resolve these geographical "imbalances" and socio-spatial "injustices."

Injustices, so the argument went, are inscribed in nature, but it should be the mission of the state to produce a just hydro-social distribution, one that would correct the errors of its divine Creator. As already suggested in chapter 3, while nature's injustices had to be undone and the "perturbed" equilibrium of the erratic hydrological cycles in Spain "restored," an intense argument unfolded between the civil engineers and the forestry engineers. While the former pursued enrolling water through reorganizing and reengineering Spain's rivers, the foresters argued for a solution that would, by means of reforestation, improve water capture and storage, diminish erosion, and optimize the water balance (Gómez Mendoza and Ortega Cantero 1987). Fueling the argument were professional antagonism and prestige, "scientific" debate, as well as the particular insertion of the civil engineers within a wider network of industrial, banking, and political networks. It would ultimately be the hydraulic and civil engineers who would carry the day: the national water economy would primarily be organized through a hydraulic techno-natural assemblage of concrete, steel, dams, reservoirs, irrigation channels, hydroelectrical power generation, and other large water infrastructures (López Ontiveros 1992, 301–302).

By the late 1930s, this socio-physical construction of water as the source of Spain's precarious condition because of its erratic temporalities and uneven spatial distribution, which can be undone through appropriate techno-natural structures, had become deeply ingrained in the elite's imaginings of Spain's condition. Social justice and national water redistribution went hand in glove in much of the period's representations. Indeed, this particular staging and mobilization of water was captured effectively by Franco and often repeated in his texts, discourses, inaugural speeches, and the like: "We are prepared to make sure that not a single drop of water is lost and that not a single injustice remains"[3] (Franco 1959b, 1).

This quest was articulated very much around the person of Franco himself. It was "his" responsibility to make sure the waters of Spain were transformed to serve the needs of its people. A commemorative inscription (see figures 5.2 and 5.3) unveiled on the occasion of the inauguration of the dam of Cenajo (Segura Basin) in 1963 summarizes this well: "Francisco Franco, Caudillo of Spain, ordered its construction. With it he dominated the turbulent waters of the Segura River so that they can fertilize patiently the thirsty land and redeem men so that they can work free of the millenarian fear of floods and droughts."[4]

The media and official public opinion makers would echo this in a plethora of muscular and exultant statements and endorsements of Franco's noble water vision. For example, Enrique Del Corral put it as follows in an editorial on the dictator's visits to one of the hydraulic projects: "It is the rhythm of the heart that gives impulse to the hearts of all the Spaniards redeemed of the thirst, of the hunger and of the sterile labor on difficult land thanks to the warm politics of Franco who spares no effort when

Paco El Rana's Wet Dream for Spain

Figure 5.2
Commemorative inscription, 1963. Dam of Cenajo.

Figure 5.3
Dam of Cenajo, Segura River in the provinces of Albacete and Murcia.

it comes to awakening Spain of its age-old morose slumber"[5] (Del Corral 1959a, 1).

Indeed, it is a widely shared vision. For example, R. Cavestany de Anduaga, minister of agriculture between 1951 and 1957, repeated the regenerationist motto that "not a drop of water that we try to get will later be lost to the sea" (Cavestany y de Anduaga 1958, 192). The propaganda machinery effectively played on this twin position of water as simultaneously the source of Spain's problems as well as the "thing" from which redemption and salvation could be wrought, the mythical harbinger of social justice. In the early postwar period, during a period of intense socioeconomic decline, severe economic and financial restrictions, continuous setbacks, persistent agitation by and prosecution of republicans of all sorts, and rapidly declining living standards, the blame for these "misfortunes" was invariably laid at the door of the "persistent drought" (*pertinaz sequía*). Franco and his supporters continuously invoked the image of enduring drought to explain the difficult socioeconomic conditions affecting most—and particularly poor—Spaniards. While the droughts of 1945 and 1949 were indeed unusually severe, average water availability over the period 1945–1955 did not significantly diverge from the long-term mean. Of course, the specter of recurrent drought, combined with the persistent argument that water infrastructures would pave the way to a wet and fertile future, was staged as one of the vital and central projects for realizing the fascist utopia (see, for example, Sabio Alcutén 1994). The extract that follows is just one among many in which Franco mobilizes water as an integral part of his politics:

> Spain pains us because of its drought, its misery, the needs of our villages and hamlets; and all this pain of Spain is redeemed with these grand national hydraulic works, with this Reservoir of the Ebro and all the others that will be created in all the basins of our rivers, embellishing the landscape and producing this golden liquid that is the basis of our independence.[6] (Francisco Bahamonde Franco, speech on the occasion of the inauguration of the Reservoir of the Ebro, August 6, 1952; cited in del Rio Cisneros 1964, 122–123)

The debate over water as well as water engineering became squarely structured around the desire to construct a nationally more equitable and just distribution of water resources by means of a grand geographical reorganization of the flows of water. The inter-river basin water transfers would become the backbone of this imagined national grid. Major engineering efforts were generated out of this view and the skeleton of this system (The Tajo-Segura transfer) was built under the Franco Regime (but only completed in 1979—see chapter 6). The draft Second National Plan would complete this system, which became defined as an "Integrated System of

National Hydraulic Equilibrium" (Sistema Integrado de Equilibrio Hidráulico Nacional [SIEHNA]) (Baltanás García 1993, 12; Ministerio de Obras Públicas y Transportes 1993a) (see chapter 7). As Martínez Gil contends, the doctrinal nucleus of the hydraulic imaginary that was forged during the long second half of the twentieth century was

> the thesis of the natural hydraulic disequilibrium in the country, with a dry Spain and another humid Spain, which had resulted in an age-old situation of deficit river basins that lack the water social demand requires, in the face of other surplus river basins where the circulating volumes of water are greater than present and future demand. This situation presents itself in a country that paradoxically has more water resources than Denmark. . . . Our treacherous torrential waters are the definitive image of a country in which the Creator has made a mistake. (Martínez Gil 1999, 110)

Entangling waters in this manner, rectifying nature's errors, and restoring a national hydraulic balance demanded the upscaling of the management and planning of water resources from the scale of the river basin to the national scale, national integration, a centralized hydraulic administration, and a strong national state with centralized and absolute power over the waters of the country, a mission Franco promised to deliver:

> This effective policy of regulation [increasing the regulated volumes of water], that has largely succeeded in correcting the uneven distribution over time of our precipitations, and to restore a regularity in the natural regimes of the rivers in order to create permanent flows where there were only torrential ones, has been the fundamental base of Spanish modern integrated hydraulic planning.[7] (Gomez De Pablos 1973b, 340)

It does not come as a surprise, of course, that toward the end of Franco's life, he was seen as the great master dam builder. The Chairman of the Spanish and International Commission on Large Dams, for example, salutes Franco in 1971, in a speech presented to him, as "the great builder of great dams and an example, unique in the world, of a statesman who creates the hydraulic foundations for the progress of his people" (Torán 1971, 314). *Paco el Rana* had indeed directed and overseen the complete socio-hydraulic revolution of his beloved fatherland. Achieving this water "activism" depended crucially on the loyal support of a series of powerful interlocked national "networks of interests" and coalitions (Melgarejo Moreno 1995, 7). They often overlapped partially, were occasionally antagonistic, and required careful massaging and "managing" within an overall Falangist program and ideology. It is these national networks of interests that supported and consolidated the Franco regime, and together with the mobilization of water, produced the assemblage that would render the socio-hydraulic edifice possible and permit it to "stand." This I shall turn to next.

Producing "Networks of Interest"

The metonymic sequence that linked H_2O's materiality with erratic rivers and the promise of rectifying the uneven and unjust distribution of their redeeming and healing qualities permitted enrolling this "vibrant matter" (Bennett 2010) into a wider flow of discourses, practices, and fantasies. Overlaid with promises of economic prosperity and socio-ecological justice, a range of different interests assembled together into a heterogeneous but relatively stable network of interests. It is precisely this network of interests that would support, sustain, and build the new techno-natural edifice upon which Spain's postwar modernization would rest. In the following pages, I shall explore how these diverse drivers and often radically different positions become entangled such that a national alliance emerged that would militantly pursue a particular vision of hydro-modernization, one that would remain uncontested for decades to come.

Forging Elite Consensus: The Falange as Ideological Glue

The *Falange Española Tradicionalista y de las Juntas de Ofensiva Nacional-Sindicalista*,[8] in short the Falange, was founded in 1933 by José Antonio Primo de Rivera, son of dictator Miguel Primo de Rivera, as an extreme right-wing and nationalist party. General Francisco Franco found in the Falange a political ally with a clear ideology at hand for his use. After a decree in 1937, the Falange and other right-wing factions were forcibly merged into one political organization. Franco became the Falange's absolute chief.

The ideological cement of the Falange, which was the unquestioned (and unquestionable) political support base of Francoism, centered on a radical rejection of both the class politics of socialism/communism and the excesses of capitalist liberalism. It cherished a firm belief in the hierarchical and organic organization of society in which everyone had his or her place, accepted it, and acted accordingly, and cultivated a hazy unifying national-cultural vision of *Hispanidad* (Spanishness). While modernizing in its economic desire, the Falange shared with other fascist and nationalist visions a conservative and mythical celebration of a heroic national Spanish past through which a particular Spanish identity was produced. Franco's Falange compared its mission with a veritable "crusade" and the heroism of *El Cid* was its great role model.[9] This eclectic and inconsistent mixture underpinned the early Franquist politics around which a series of networks of social actors galvanized. Its vision is summarized eloquently in the Labor Code (decree of March 9, 1938):[10]

Renewing the Catholic Tradition of social justice and high human signification that informed our Imperial legislation, the National State is a totalitarian instrument in the service of the integrity of the fatherland, and as such represents a reaction against liberal capitalism and Marxist materialism, and undertakes the task of carrying out—with military flair, constructive and deeply religious—the pending Revolution of Spain and that must return to the Spaniards, for once and for ever, the Fatherland, Bread and Justice.[11] (cited in Tranche and Sánchez-Biosca 2002, 187)

The socioeconomic alliances that were forged skillfully by Franco generated a labyrinthine network of power relations that supported the regime and assured its longevity. The Civil War had of course eliminated, either physically or through imprisonment or exile, the most activist parts of the opposition movements, both liberal and socialist-communist. In the process, Franco had assured himself of the loyalty of many royalists, the large landowners, the Catholic hierarchy, the military cadres, significant parts of the national industrial bourgeoisie, nationalists, and national-syndicalists. The strong linkages between state and economy would cement a corporatist state structure that could count on an endogenous capitalist sector, whose success and profit was closely tied up with the state's investment flows.[12] In the present context, I shall concentrate on those networks that have been somewhat neglected in the literature (in contrast to the role of the Falange, the military, or the Catholic Church), but nonetheless proved vital for helping to secure the geographical revolutions that were such an integral part of Franco's vision. In doing so, I contest the dominant historiographical view that sees experts, engineers, science and technology, and "the masses" as operating under the dominance of the fascist leader or centralized party. I concur with Lino Camprubí's argument that these actors are "active participants within the regime" that play an "active role in producing these political mandates and ideologies" (Camprubí 2012, 23). I focus on the key ideologues and practitioners that provided the technical, scientific, and discursive support that would build and maintain—both materially and symbolically—the expanding networks of dams, pipes, hydro-machinery, and irrigation systems, namely the large landowners, the electricians, the engineers, the geographers, and the media.

Water for the *Latifundistas*

While the rise of popular movements early in the century had raised the problem of peasant laborers and their right to land and water, the outcome of the Civil War solidified the interests of the large landowners, particularly, but not exclusively in Southern Spain (Ortí 1994, 243; Bernal 1990). With the victory of the fascists and the end of the Second Republic, hydraulic politics took an important new turn. Of course, there was considerable continuity in terms of the technocratic-engineering vision of hydropolitics, but the socially reformist agenda that had been an integral part of the republican view was radically altered. In particular, the relationship between land and social reform on the one hand and hydraulic infrastructure on the other was broken, eliminating everything that could be a threat to the interests of large landowners (del Moral Ituarte 1991, 508). The link between irrigation planning and peasant agriculture was radically dissolved, restoring the hegemony of the landowners (del Moral Ituarte 1999, 186–187) while still paying lip service to considerations of social justice. Attempts at significant land redistribution stopped. Under Franco, the link between water policies and land reform was severed. A newly established institution, *Instituto Nacional de Colonización* (INC) now managed the latter. This was set up to provide land to landless peasants and became a great propagandistic tool, but achieved relatively little. Ultimately, the INC acquired only 149,358 hectares of irrigated land and settled 24,047 colonists on these lands between 1939 and 1975. Another 323,385 hectares of nonirrigated land was acquired, which was offered to a total of 23,773 peasants (Ortega 1975, 240). While internal colonization propaganda would remain powerful in sustaining the official Franquist desire to redistribute land and facilitate the expansion of a class of small landowners, the resettlement of landless peasant in newly irrigated areas was minimal in comparison to the extraordinary state-financed irrigation of the land of large landowners.

An estimated total of 1.635 million hectares of newly or improved irrigated lands were serviced by the state (Ortega 1975, 223). Indeed, the earlier socially motivated hydraulic regenerationism could easily be transformed into an ultra-protectionism of the *latifundistas* by the dictatorial regime (Acosta Bono et al. 2004, 112; Ortí 1984). As Nicolás Sánchez-Albornoz maintains, "The land owners received the double gain of both an increase in production from their irrigated lands as well as the re-valorization of their lands, without much counterpart other than to support the regime, something they unfailingly offered" (Sánchez-Albornoz 2004, xxv). The systematic opposition of the large estate owners to state-led hydraulic interventions (precisely because it was initially often implicitly or explicitly linked to land reform) melted quickly as not only the leftist threat to their position dwindled but as they also gained an extraordinary increase of the value of their lands as the state basically funded turning the semi-desert into fertile and high-yielding fields. Letting the desert bloom became an extremely profitable accumulation strategy. Indeed, while the state covered the cost of infrastructure, the landowners reaped the benefits, with an estimated 1,200–2,000 percent improvement of their economic return (Bernal 2004, xxxvi). The production of land rent and its private appropriation through the mobilization of collective resources, a process that orchestrated extraordinary transfers of value to the landowners, would be one of the disavowed but very real hard kernels of Franco's hydraulic politics. This twin condition of eliminating left-winged forces while generating extraordinary wealth for the large landowners turned most of the latter, not surprisingly, into one of the most solid social and political pillars on which Franco's political and sociocultural edifice would rest.

While internal colonialism (Ortega 1975) and the "social land problem" would still be rhetorically mobilized throughout the postwar period, Franco's hydro-politics has to be characterized as an agricultural "counter-reform" that guaranteed the long-term stability of the *latifundia* system (Martínez Alier 1968). However, while the discursive-ideological image would continue to stress the relationship between hydraulic works, the plight of the peasants and the development of irrigated agriculture, the production of hydroelectrical energy, and the support of the electrical industry would de facto become the central concern.

Hydro-Electrifying Spain

Despite the recurrent rhetorical attention paid to the irrigation "mission" of the state's hydraulic project and the emphasis on internal colonization, the formation of smallholding irrigation farming, and integrated planning (like the *Plan Badajoz*),[13] this discursive ploy served primarily as grist for the propaganda mill and for the rhetorical arsenal mobilized to legitimize the implementation of grand hydraulic infrastructures (Díaz-Marta Pinilla [1969] 1997, 73). One of the great efforts—but one largely neglected in the abundant literature on irrigation and colonization—went into securing the necessary energy resources for the self-reliant development of Spain (Gomez De Pablos 1973b, 338). James Simpson, for example, argues that hydroelectrical power generation constituted the prime aim of the policy, with the extension of irrigation as a byproduct: "Although the principal goal of such development [massive construction of reservoirs] was hydroelectric power, the area of irrigated land also increased by 600,000 hectares (41 percent) between 1950 and 1965" (Simpson 1995, 261).

There was indeed a significant difference between rhetoric and reality. For example, between 1940 and 1963, 322 dams were constructed, of which only 132 had irrigation as their principal goal (Melgarejo Moreno

2000, 302; Barciela López and López Ortiz 2003, 65). Until the late 1950s, more than 75 percent of the energy needs of Spain were secured by hydroelectrical power. Between 1939 and 1957, installed hydroelectric capacity increased from 1,400 Mw to 5,200 Mw, generating a total production of 2,844 million kWh (kilowatt hours) in 1939, increasing to 18,790 million kWh in 1957. Here too, the bulk of the expansion took place after the mid-1950s, representing a total value of approximately $458 million (in 1957 parity terms) (Garrido Moyron 1957). After 1964, the relationship between irrigation and hydraulic works was further severed in favor of hydroelectrical developments. Of the dams constructed between 1964 and 1977, only 96 (38.2 percent) were destined for irrigation purposes, while 57.6 percent of the created capacity was earmarked for energy generation. In addition, twenty-nine hyper-dams (of more than 1,000,000 m^3) were constructed; many of which were also vital for the regulation of electricity production (Vera Rebollo 1995: 313). By the end of Franco's rule, total installed energy capacity was over 25,000 MW (megawatts) and production had reached 82,000 GWh (gigawatt hours) (Antolín Fargas 1997, 202). Although the contribution of hydroelectricity had fallen from 78 percent in 1949 to a still significant 46.9 percent in 1975, hydro-energy was absolutely vital for Spain's modernization (see figure 5.4).

Together with improving and expanding urban water supply, the hydroelectrical development of Spain became the real mission. Both the geographical location of newly constructed dams (primarily in the North) and the extraordinary expansion of the hydroelectrical grid testify to the entanglement of hydraulic interventions, private economic interests, and national energy provision. Moreover, the rapid expansion of cities like Madrid, Bilbao, and Barcelona, and their dismal sanitary conditions, would become a major problem and obstacle. Together with the tourist boom that started in the early 1960s, dealing with the urban water supply conundrum would begin to absorb more resources and become a key priority. While the hegemonic discursive focus remained squarely fixed on expanding irrigation, improving agriculture, and containing flood risks, the urban condition and the exponentially growing demand for energy drove much of the actual hydraulic infrastructure developments.[14]

While private initiative dominated in the energy sector, the tutelage and coordination of the state was considered vital. And yet again, the engineering of the country's national waters was the pivotal means through which to achieve this. The Conde de Guadalhorce, minister of development under the dictatorship of Primo de Rivera, and later great supporter of Franco, insisted on the primacy of hydroelectric power generation for

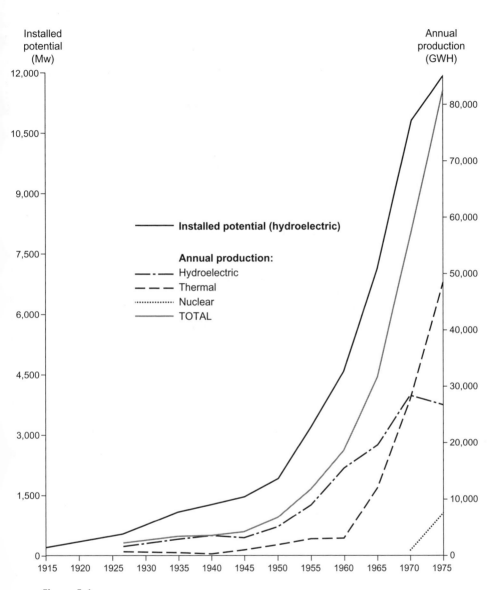

Figure 5.4
Electrical potential and production, 1915–1975.
Sources: García Alonso and Iranzo Martín 1988, 344–345; Vallarino Cánovas del Castillo and Cuesta Diego 1999, 2000; and Ministerio de Medio Ambiente 1998, 311.

the development of Spain and the need for collective and integrated state action:

> It is not the waters that run through the mountain ranges and slopes directly to the sea, however beautiful they are, that are the most profitable, but those that are dominated, those that are supplied to common causes; these are the ones that provide wealth for industry and for agriculture. And this has to be applied not only to the hydro-electrical issue, but also to all the manifestations of industry and to the progress of Spain. . . . The hydraulic wealth that is lost is 21,000 million cubic meters to the Mediterranean and 26,000 million to the Atlantic, and there are very [water] poor areas and areas with abundance, and we must research the transfer of water from some zones to others. That is to say, our hydrographic system does not permit to divide the country in regions and obliges us to consider the totality, since otherwise we would not have equilibrium in the country.[15] (Guadalhorce 1950, 509)

The electricity production sector was closely allied with the "network of interests" that sustained the fascist polity (Núñez 2003). The immediate post-Civil War period saw an intense process of vertical and horizontal integration of electricity companies and an interlacing of the state with these oligopolistically organized energy companies (Buesa 1986; Antolín 1999). The geographical integration of capital and organizational structures was paralleled by a national territorial integration of the electricity network through the production of a national high-voltage grid (Puente Diaz 1949). Until the mid-1950s, energy shortages were a continuous issue, and invariably blamed on the *pertinaz sequía* (perennial drought) (Sudriá 1997). With the exception of the exceptionally dry years of 1945 and 1949, however, the shortages were undoubtedly the result of insufficient capacity and investment, combined with shortages of heavy equipment after the Civil War, and of the autarchic and isolationist position of Spain (Sudriá i Triay 1990). Nevertheless, the "perennial drought," mobilizing the tried tactic of blaming a hostile and uncooperative nature, proved a powerful discursive vehicle that not only shifted the blame onto nature, but also provided the glue through which the hydraulic mission of Spain could be legitimized and extended to include urban-industrial development and the vital provision of homegrown energy.

The acute energy shortages during the autarchic period of development, however, did not prevent the electricity companies (and their banking allies) to be among the most profitable businesses in the country. The state's policies and interventions generated a significant transfer of state capital from the public to the private sector, either indirectly through major hydraulic works that regulated the erratic flows of the rivers to permit continuous production or directly through subsidies, cheap loans, and cross-capitalizations (Antolín Fargas 1997). Although private companies indeed invested in the construction of dams and electrical generation under concession from the state, their contributions covered only a small part of the total cost of regulating the entire river flow. For example, massive regulatory dams were constructed by the state upstream of rivers. This, in turn, permitted the electricity companies to build their own infrastructures downstream with a much more regulated and controlled continuous flow of water. Indeed, uninterrupted hydroelectrical production was predicated on the regulation of the flow of the entire river, and this required massive state-funded infrastructure to permit the private dams to function properly. Some of the largest energy oligopolies were created in Spain during this period, with the public works and industrial policy administrations as their main protagonists (Núñez 1995). A symbiotic relationship developed between the state and the energy producers, something openly presented as a mutually beneficial undertaking (see, for example, Vicens Gomez-Tortosa 1961, 438–439).

Most of the investment for hydraulic works went to the North where the greatest potential for hydroelectrical power generation was located. As table 5.1 suggests, the largest number of dams during the 1950s and 1960s were constructed in the North, while the most urgent irrigation needs were, of course, in the South of the country. The regenerationist discourse of agricultural modernization through irrigation played a powerful ideological role

Table 5.1
Dams constructed by river basin

River basin	Before 1940	1941–1955	1956–1970	1971–1980	1981–1986	TOTAL
Norte	11	22	67	25	11	136
Duero	9	15	29	5	3	64
Tajo	37	21	51	52	20	184
Guadiana	27	3	15	28	15	85
Guadalquivir	15	11	28	16	3	86
Sur	8	0	2	8	7	25
Segura	8	1	7	7	1	24
Jucar	13	9	16	4	0	42
Ebro	59	24	61	8	16	168
Total	187	106	276	169	76	814

Source: Ministerio de Obras Públicas y Urbanismo 1990, 33–34.

to legitimize large-scale hydraulic engineering, yet a significant share of the actual works were directly related to increasing energy production. This was one of the clearest ways through which the original hydraulic dreams of the early modernizers was perverted to serve the particular interests of the oligarchic elites that fused around the fascist national project.

A Faustian Pact: the Corps of Engineers at the Core of Franco's Project

The great transformation envisaged by the regime depended on the loyal support of the Spanish Corps of Engineers. The civil engineers were indeed the key protagonists of the preparation and implementation of the hydropolitical agenda and guaranteed continuity in the execution of Franco's dream (Gil Olcina 2003, 56). Their technical expertise and engineering projects intertwined closely with the regime's political and social vision, and they became active participants in assuring its cohesion and longevity (Camprubí 2014). The engineers were, of course, traditionally closely associated with the state apparatus, but, under Franco, the quest for a newly manufactured national hydraulic geography by means of the "rebirth of public works and the success of an efficient hydraulic politics" (Sánchez Rey 2003, 26) propelled the engineering fraternity (they were all men) to the forefront of Spain's fascist modernizing project. The pages of the *Revista de Obras Públicas* (ROP) reflected the views and visions of the Corps in relationship to the social, political, and engineering themes of the time (see Songel González 2003, 83).

The engineers, as much as any other segment of society, were politically split during the Civil War. After the beginning of the Civil War in 1936, the engineering school and its associated *Revista* were taken over for a short period of time by members of the leftist Union of Architecture and Engineering of the *Union General de Trabajadores* and in an editorial of August 15, 1936 (ROP 1936), entitled "Establishing Positions," they called for closed ranks in fighting off the fascist enemy and for building a modern and civilized Spain, while recognizing that the engineering fraternity was generally perceived by the public as "lieutenants of the reactionary companies, exploiters of workers, or entrenched bureaucrats" (ROP 1936, 1; Sáenz Ridruejo 2003, 12):

Today, we must devote all our forces, without either lack of enthusiasm or cowardice, to the service of Spain and the Spanish economy, which is in a critical moment of destruction because of the fascist treason. To raise Spain, we need all men of good will . . . Today it is about the defeat of the enemy. Tomorrow, the hour will sound to construct a Spain that stands as a model of civilized countries.[16] (ROP 1936, 1)

Only six issues, reduced in size and badly distributed, appeared in 1936 under the editorship of the union-controlled *Revista*. However, when after the Civil War "official" publication resumed on March 1, 1940, the ranks had closed solidly around the fascist triumph. The union-published issues of the journal were repudiated and their editors were described as "red bandits." Numbering just continued from the last "official" issue of July 1936. Indeed, by early 1940, the engineering profession had fully embraced the new "visionary" politics embodied by the Franco regime. Never before had the engineers endorsed and unequivocally supported a political regime with such vigor and unmitigated enthusiasm. A special issue, dedicated to the "Spanish Crusade—1936–1939," was published, with a portrait of the Caudillo on its front page, subtitled "FRANCO! FRANCO! FRANCO!" In its opening pages, homage was paid to the heroic colleagues that had fought and died during the "brilliant campaign of liberation" on the side of the Generalissimo (ROP 1940, 2003a, 53). The new regime was celebrated in exuberant style. The Corps pledged its unconditional support to the nationalist cause and embraced the Falangist vision in its opening editorial:

> The ROP resumes today its publication, after an interruption of almost four years. From the beginning of the Glorious National Movement [the Franco uprising], all the sympathies of the Editorial Committee were on the side of the Government that came to restore the eternal Spanish tradition. The reds knew and distrusted this, perhaps expecting an attachment to its cause either enthusiastically or explicitly or perhaps under the threat of terror. And as this did not happened as they wished, a group of undesirables, pistols in their belt, broke into our offices a few days after the August 1936 issue had been released, and seized the magazine on behalf of the National Union of Architecture and Engineering of the General Union of Workers. They even tried to have the editorial board to continue its works under its improvised leadership, but none of its members were prepared to collaborate under such conditions. And a few days later, under a new Editorial Board, they continued to publish a few issues, using the original copy prepared by us; but when this was exhausted, as happened after three months, its efforts ended.[17] (ROP 1940, 1)

To take away any remaining doubt as to the allegiance of the engineers to the new regime and its causes, the issue also reproduced a speech by Civil Engineer Tomás García-Diego de la Huerga delivered on October 17, 1937 during a ceremony of pledging allegiance to the flag. It was a sign of the ideological principles to be followed from then onward. In his eulogy, he stated the need "to recover Spain's imperial vocation for which it proposes to enhance our religious faith, making it the basis of our activities and honor all racial virtues. Against the false dogmas of rotting democracies, the mottos of our Golden Age, now already embodied by the Generalissimo. Against freedom, service. Against equality, hierarchy. To overcome fraternity, a brotherhood which presupposes the common paternity of God"[18] (ROP 1940, 5; also cited in Aguiló Alonso 2003, 78).

The engineering community (at least most of those who stayed behind and were not exiled, jailed, killed, or had emigrated) would put their collective efforts into modernizing the country within the collective enterprise shaped by the new regime.[19] Public Works became one of the pillars of the regime—in Franco's words, "an excellent means of protection and stimulator of its prosperity" (ROP 1940, 2). Indeed, "the Corps of Engineers constituted consequently one of the most solid supports of the policies of the new regime" (Songel González 2003, 84). In the first Franco government, two civil engineers were appointed to ministerial posts: Alfonso Peña Boeuf for Public Works and Petro González Bueno for Syndicalist Organization and Action. During the following forty-five years, no explicit political statements were made by the Corps of Engineers, but the journal filled many of its pages with countless celebratory articles extolling the virtues of dam constructions, recounting tirelessly and in endlessly reiterated technical detail the achievements of newly built dams, and providing annual summaries of dam constructions and progress reports on the execution of grand hydraulic projects. Furthermore, it provided detailed celebratory and hagiographic reports of Franco or other government dignitaries visiting and inaugurating major water projects, extolling each time in hyperbolic terms the rapturous reception that Franco unfailingly received from the local people during each inaugural visit, praising the great progress the Franco-engineering alliance was making in developing the country and raising it from the doldrums of both drought and poverty.

For example, in the issue of June 1961, in a self-congratulatory hymn to the virtues of the Spanish hydraulic engineers, José Luis Mendoza Gimeno gave a poetic evocation of how these engineers, serving the national(ist) cause, possess a

"hydraulic sensitivity," a sort of sixth sense, that permits to intuit the comportment of water in its movements. . . to be a good hydraulic engineer one has to know, see, hear, touch the water with the eyes, the ears, the hands. . . with a vibrant soul. . . . It is not a surprise that Spain had and has the most magnificent hydraulic engineers. . . because Spain is the country of the imagination par excellence. . . . If history feeds the racial imagination of the Spanish, Geography and the climate do not contribute to a lesser degree. . . . The burning soil that gives pleasure and scorches contributes to the scarcity and irregularity of water on the Peninsula. . . . These two characteristics combined produce such deficits of water that a determined struggle of the Spaniards to domesticate it extensively and continuously over the whole Nation [has been fought]. . . . The biblical phrase "by your labor thou shall know you" applies to our hydraulic engineers. . . . To all those, those who have passed and those that today continue their work so brilliantly, the gratitude of Spain is theirs.[20] (Mendoza Gimeno 1961, 364–367)

In the same issue, a long list of built dams and their characteristics was published, together with the list of all the companies that have contracted hydraulic works (see table 5.2). The network between state-, engineering-, and construction companies ran through this self-congratulatory special issue. The list of companies is telling in itself. All but one of them had been set up during the dictatorship and worked almost exclusively for the Public Works program of the state. In addition to providing a detailed list of their technical capacities and engineering achievements, the data also reveal the importance of hydraulic works for employment. These nine companies alone provided jobs for 358 engineers and had a total direct work force of more than 48,000. This sector alone therefore made an important contribution to the economy, particularly during the early period when jobs were a scarce commodity.

In December 1961, in an equally aggrandizing review of "20 years of hydraulic politics," engineer Francisco Casares celebrated that now "the rivers are controlled and managed, and we obtain from them the maximum return" (Casares 1961, 911). In 1971, the chairman of the Spanish and International Committee for Large Dams saluted, in front of Franco, the engineering contribution "of raising the flag, the Spanish flag of grand dams, sustained by 20 years of glorious history, and adorned by the ribbons of the 360 battles [i.e., newly constructed dams] that your Excellency succeeded in winning" (Torán 1971, 315). This self-congratulatory, state-centered and Franco-adoring style of the *Revista* would continue for another seventeen years, together with the uncritical defense of the major public works programs and relentless support for the further consolidation and final completion of the hydraulic project for Spain.[21] Only from 1992 onward did a more critical and socially engaging style gradually began to emerge (Nárdiz Ortiz 2003, 104). Although the "steel and concrete fever" remained an obsessive theme, fed by the drive to restore Spain's hydraulic equilibrium by constructing more, the first voices of dissent began to be raised. Yet, the "hydraulic structuralism" of the Franco years would remain the guiding force of hydraulic policy and planning long after the end of the dictatorship (Arrojo and Naredo 1997).

Table 5.2
The nine dam construction companies and their workforce in Spain, 1961

Name of company	Active since	Engineers	Other technical staff	Administrative staff	Manual workers (average)	TOTAL
Agroman Empresa Constructora S.A.	1927	101	300	394	10,207	11,002
Termac Empresa Constructora, S.A.	1943	12	51	78	2,072	2,213
Cimentaciones Especiales S.A.	1936	16	33	47	888	984
Agrupación para Estudios y Proyectos de Obras	1953	60	267	716	6,335	7,378
Dragados y Construcciones	1941	80	169	379	10,944	11,572
Empresa Auxiliar de la Industria S.A. (AUXINI)	1945	52	187	117	7,300	7,656
Obras y Construcciones Industriales S.A.	1942	15	45	60	3,952	4,072
San Roman S.A.	1943	4	12	29	197	242
Helma S.A. Empresa Constructora	1952	18	100	92	2,780	2,990
TOTAL		358	1,164	1,912	44,675	48,109

Sources: NN 1961, annex; Ministerio de Obras Públicas y Urbanismo 1990, 33–34.

The Patriotic Geographers of a Unified Spain

While Spain's geography was undergoing its most momentous and spectacular socio-physical transformations in history, academic and professional geographers remained symptomatically silent about these changes. Instead, geographers chose to play and geography played a pivotal role in creating and sustaining a patriotic vision of a mythical unitary Spain; a vision that complemented, at a representational/discursive and imaginary plane, the physical-geographical unification pursued by Franco's engineers. Indeed, in the transition from the Civil War to the postwar Falangist "stabilization," a radical transformation swept through academic geography. A "fascist" geopolitical discourse was thriving during the war, one that celebrated the nationalist-imperialist project with which Franco had aligned himself (see Vicens Vives 1940; 1941a,b; 1981) and was modeled after Nazi geopolitics and its materialist-ideological constructions of race, nation, and empire. The core themes of this "geographical" analysis focused on the indisputable unity of Spain as a vital (if not vitalist) national living space and insisted on its manifest "imperial" destiny. The latter would not necessarily take a territorial form predicated upon military conquest and geopolitical control (as this was firmly out of reach of Spanish military capacity) but one that centered on the celebration, cultivation, and spread of *Hispanidad* (Spanishness), particularly to Latin America. This radical fascist geopolitics paralleled much of contemporary military geopolitical thought (Reguera Rodríguez 1991).

However, after the war, the discredited principles that had galvanized fascist geopolitics no longer served the interests of the consolidating fascist regime, which preferred to see a much more depoliticized and less radical (at least explicitly) geography emerge (Gómez Mendoza 1997), one "that should lead to the aggrandizement of the Fatherland," as the director of the Institute of Geography Juan Sebastián Elcano within the High Council for Scientific Research (HCSR) and editor of the newly established journal *Estudios Geográficos* put it in one of the early issues in which he called for a series of urgent reforms in geographical education (Bullón 1941, 676). His vision advocated the promotion of a depoliticized geography regime and curriculum in both secondary and higher education (Reguera Rodríguez 1991, 52). J. L. Asían Peña summarized the sort of geography the Institute promoted as one "that widens the horizon of our knowledge of the Universe, the World, and the Fatherland, and that also increases our Patriotism and our love for the Highest Creator" (Asían Peña 1941, 10).

In practice, geographical research became the study of regional and local landscapes that would forge a mythical unity of Spain through celebration of its diversity. Taking its cue mainly from French possibilism and regional geography, Spanish geography was considered and practiced as a physical-cultural assemblage of diverse regions, which nevertheless shared a set of common characteristics and a national unity. Politics as such disappeared completely from the geographical agenda, yet the political-ideological role of geography remained, as ever, significant. This ideological edifice was further consolidated by the official "national-Catholic" doctrine that permeated the educational system, the censure of books that did not "respond to the principles of religion and Christian morality and exalt patriotism" (Sánchez Pérez 1981, 10); by laws that repressed masonry, and communism, or raised regionalist aspirations or concerns, and the strict state control over access to teaching and academic posts, including the exiling and depuration of those considered hostile to the regime. For Franco, national unity was a sacred and mystic ideal. Geographers came to embody and express this vision of Spain defined as "the Fatherland as a spiritual, social and historical unity" and Catholicism as the undisputable "crucible of nationality" (Franco, June 24, 1938, cited in García Alvarez 2002, 354; see also Richards 1999). *Hispanidad* was as much forged by a shared culture and history as by the unity of its diverse regional geographies and *genres de vie*. The ideologically driven nature of depoliticized university education became enshrined in the 1943 Law on University Education:

> The fall of the monarchy further precipitated the catastrophe of our centers of culture, and the Republic launched the University on the slope of its annihilation and *de-españolización*, to the extent that from this sprang the most monstrous national negations. . . . The Act, in all its provisions and articles, calls for the faithful service of the University to the ideals of the Falange, inspiring the state, and vibrates to the rhythm of the imperative and the style of the heroic generation that knew how to die for a better fatherland. . . . In undertaking this educational and cultural transformation, the most fecund and urgent slogan of the National Revolution is carried out, one required by the blood of those who knew how to die in the line of duty and by the noble passion of those who now want to also serve Spain's supreme destination with their life.[22] (José Ibáñez Martin, Minister of Education, 1943, cited in Sánchez Pérez 1981, 12)

History and geography as academic subjects were indeed considered important as carriers of this unitary nationalist ideology and as supporters of a patriotic spirit (Capel 1976).[23] Within the HCSR, geography received generous support from the state, which of course controlled educational programs strictly and made sure to eliminate those geographical themes that did not rhyme well with official politics. This state-led educational propaganda system was paralleled by a growing self-censorship of intellectuals whose professional position depended strictly on their allegiance to official doctrine (García Alvarez 2002). Being doubly blind, both to the geopolitical tactics of Franco's regime and to the actual geographical transformations that were actively producing a new Spanish landscape, geography in effect served as a legitimizing veil for the "patriotic" and unifying mission of Franco's regime. Indeed, only scant attention was paid to the techno-natural transformation of Spain's waterscape. While the actual participation of geography and geographers in the planning and implementation of geographical projects was nonexistent (Capel 1976, 30), the ideological role of geography and the active enrolment of geographers in the "network of interests" (Sanchez Recio 2003, 13) that supported Francoism was nevertheless significant. Naturally, the production of consent in academia was paralleled by the discursive production of the virtues of Franco's regime in the popular media. The latter would indeed prove to be vital in securing and maintaining consent. This is what we shall turn to next.

Galvanizing the Nation: Propaganda and Hydraulic Works

Like in Germany and Italy (see Caprotti 2004, 2005a,b), in Franco's Spain sophisticated propaganda machinery was quickly put in place after the fascist victory. As soon as Franco had consolidated his position as head of state, an official state institute for documentary filmmaking called NO-DO (*Noticiario Documentales Cinematográficos*) was established in addition to imposing the tried tactic of controlling and censuring the print media. Grafted on the popular cultural success of cinema, NO-DO produced news and general interest film reels that were required to be shown in the country's cinemas and also were sent abroad for screening.[24] This highly subsidized propaganda instrument served to celebrate the regime personalized by Franco, to animate the enthusiasm of the people for the regime's efforts, to eulogize the virtues of Spanish traditional cultural values, and to mythologize the "crusade" for a reinvigorated, conservative, and Catholic Spain, one that was nevertheless prospering and modernizing (Rodríguez 1999). Between 1943 and 1981, when NO-DO was finally abolished, about 4,012 documentary reels were produced.[25] Until television entered the scene in 1956, NO-DO was the main cinematographic information source available to the wider public. In analyzing the content of NO-DO's reels, Rodríguez points out that "above all, inaugurations" filled the screens (Rodríguez

1999, 223–224). A symbiotic relationship was systematically instilled between Franco and the great national public hydroelectrical and irrigation mission. Franco personally inaugurated hundreds of dams and other hydraulic works. Each time, the event was widely publicized by NO-DO, and covered in great hagiographic detail in magazines and newspapers as well as in the specialist engineering and professional journals. On each occasion, Franco was welcomed by the "grateful and admiring masses" as "the victorious Caudillo of Spain," celebrating the "enormous social works of the country" and detailing the "great technical achievements" of the country that were illustrated by the newly inaugurated "transcendental" hydraulic public projects. The following off-screen voiceover from the reel of March 8, 1943 (reel 10) is symptomatic of such exaggerated exaltation: "The Caudillo of Spain, who during the hours of the war led our troops to Victory, is also the soul of this labor of reconstruction, with which Spain heals its wounded, saving Spain from all the difficulties that the current international circumstances pose against her."[26]

On occasion of the provisional inauguration of the Dam of the Generalissimo, the offscreen commentator of the filmed record narrates how the waters of the reservoir will "soon become gold and assure the supply of water to the lands of the Levant and prevent that the crops will be lost" (reel 229, May 26, 1947). Each new feat is welcomed as a great moment in the progress of the country, detailing in every excruciatingly boring detail the technical specificities of the inaugurated "great labor of the nation."

Inaugurations, the endlessly repeated images of the regime's activities and achievements, and the unconditional support the invariably grateful and boisterous popular masses showed to their leader fused together in NO-DO's reels. They became like a monographic documentary, a festival of laudatory images and commentaries. The image of inaugurated dams was the most iconic image of Franco—as mentioned previously earning him the popular nickname of *Paco el Rana*—who oversees the new hydraulic landscape, listens to the adulations of his entourage, and receives graciously the ovations of the masses in whose name and for whose future the works were undertaken. These frequent visits suggested a Franco that was close to the pulse of the nation, sensitive to the needs of "his" people, and attentive to the transformations taking place in the country, while emphasizing the transcendental role of himself and his state in rousing and producing the creative energies that made these transformations possible (Tranche and Sánchez-Biosca 2002, 215). The inauguration sites became emblematic representational spaces, symbolically and materially conveyed through the newsreels (see figure 5.5a, b). They were presented as geographical symbols of and material referents to the unmitigated success of the Francoist project, emblematically embodying a technocratic developmentalism while evoking the beauty, unity, and traditions of the Spanish landscape. The images celebrated the solidarist, spiritual, and moral values of traditional Spain, the tenacity of its workers, the voluntarism of its elites, the power of the regime, and the virtues of technical modernization (see Tranche and Sánchez-Biosca 2002). The aesthetics of a produced sublime socionatural environment fused with the myth of development that, combined, enchanted participants and viewers alike. With the advent of television in 1955, Franco was deeply concerned about how the ease of traveling of information might "enable the outside air to penetrate our windows, contaminating the purity of our environment" (Franco 1960, 122); he would not allow "its vitiated air to invade us" (Franco 1964, 406). Indeed, TV broadcasting was kept under close censorship control and inserted into the arsenal of instruments to celebrate the virtues of the Spanish people and its Caudillo (Palacio 2005).

The newspapers were of course equally marshaled to espouse the virtues of the regime and its achievement. On a daily basis, the press would report ecstatically about yet another great speech from the Caudillo as yet another sublime achievement was inaugurated. The quotation that follows offers a sample from an average speech by Franco given on July 1, 1959, inaugurating one of these "transcendental grand hydraulic projects":

We have come to visit your province, to inaugurate various transcendental works, to inaugurate a series of them, and with this to satisfy the thirst of your fields, to regulate your irrigations, which shall increase your welfare and multiply production. . . . We need to irrigate our fields, satisfy the thirst of so many lands. . . . The whole of Spain has to be redeemed, sealing the brotherhood between the land and the men of Spain.[27] (Franco 1959a, 1)

And the reproduction of the speech in the newspaper is invariably accompanied by equally exalted, triumphant, and jubilant commentary by journalists:

We are tightening the siege against the misery and the thirst of Spanish men and lands, because against the old sterility of the rivers, Franco's program built walls of steel to bring light, and veins of concrete to canalize the water and to take it to fulfill its irrigation mission, redeeming the old peasant thirst. And the peasants, full of joy, offer Franco today the expression of their unlimited gratitude and their unconditional allegiance.[28] (Del Corral 1959b, 1–2)

a

b

Figure 5.5
Media imagines of Franco inaugurating hydraulic infrastructure. (a) "El Caudillo inaugurates a dam on the river Eume (La Coruña)" (September 12, 1960); (b) "El Caudillo inaugurates the hydroelectrical dam of Villariño (Salamanca)" (November 14, 1970).
Sources: (a) Ministerio de Educación, Cultura y Deporte, España, Archivo General de la Administración, Fondo MCSE, signatura 33-03305-00008-001. (b) Ministerio de Educación, Cultura y Deporte, España, Archivo General de la Administración Fondo MCSE, signatura 33-03304-00010-003.

"There is no other cry that rises from the whole of Spain: Water!," the *ABC* newspaper insisted (July 4, 1969, 65) and the article continues to elaborate how "the inhabitants of the Spanish wastelands demand dams, canals, irrigation," something that "Franco knows very well" as he travels throughout the country and inaugurates the public works that maximize the use and mobilization of water that otherwise would have been lost. A few days later, on a similar occasion, the same journalist reports: "It is the rhythm of the heart that gives impulse to the hearts of all the Spaniards redeemed of thirst, hunger and sterile work on difficult land thanks to the warm-hearted policies of Franco who spares no effort when it comes to awakening Spain of its old moroseness"[29] (Del Corral 1959a, 1).

This heroic mission, thus visualized and narrated, was spread throughout the country, galvanizing the hearts and mind of the Spanish and urging them, like the grateful peasants, to embrace the remaking of the Fatherland. This consequently led to a widening of the networks of interests that would maintain loyalty to the fascist modernizing cause.

Hydro-Modern Hegemony and the End of the Spanish Labyrinth

Gerald Brenan gave his classical account of the social and political background of the Spanish Civil War the telling title *The Spanish Labyrinth* (Brenan 1943). Indeed, the pre-war period was one of tumultuous political and social life in Spain and of intense conflict along a range of different axes. The outcome of the Civil War and the fascist victory would end all this and bring the nation's destiny under unified authoritarian control. Opponents had fled, were exiled, in prison, or had died in the trenches of the Civil War or at the hands of Franco's executioners. At the same time, the regime's coherence was predicated upon forging often-uneasy alliances in a more or less cohesive network of diverse interests, aspirations, and ideological visions. In this chapter, we showed how water was discursively and materially enrolled in a relational assemblage that fused together around a hydro-social modernizing mission. This consensual configuration drew on, while also transforming, the mythical powers of water that the regenerationists had already invoked as the basis for Spain's redemption. However, during the fascist period, the raging debates over what to do and how and when to do it were silenced and the technocratic desire for its implementation was all that was left. It is precisely this uncontested hegemony that would lead to what some authors have called Spain's hydro-structuralism, in other words, the view that large-scale and state-led grand hydraulic works aimed at maximizing the availability of water at the right time in the right places were unquestionably the necessary trajectory for hydro-modernization. Indeed, the alliance that underpinned this hegemonic view would continue long after Franco's death and still infuses much of the current water debates in Spain.

However, while this chapter detailed the various circuits through which water was enrolled, the actual material transformation of Spain's hydro-social landscape would be a long and torturous process, one that would be only partially achieved by the end of the 1970s. Several decades and some major political-economic and geopolitical changes were required in order for *Paco el Rana's* imaginary to turn into geographical reality. And that is what we shall turn to next.

6 Welcome Mr. Marshall!

Almost a quarter of a century has passed since a great and brave people begun, under the leadership of a soldier-statesman, its heroic and successful campaign to repel for ever all the roots of communism. I am referring to Spain, our friend and ally, and its leader and Chief of State, Generalissimo Franco.
—Senator Henry Styles Studes Bridges, U.S. Senate, July 18, 1959, *Diario ABC*, July 18, 1959

In 1953, Luis García Berlanga, one of Spain's most venerated movie directors, released *¡Bienvenido, Mr Marshall!* (Welcome, Mr. Marshall!), a wry comedy that caught the wrath of Franco's censors, but became an instant Spanish cult classic (see figure 6.1). The title of the movie refers of course to the Marshall Plan assistance that most Western European countries had received from the United States to help revive their devastated postwar economies while providing a market for U.S. exports. However, fascist Spain had been excluded from this program precisely because of its authoritarianism and wartime association with the Axis powers. Coscripted by Juan Antonio Bardem, uncle of actor Javier Bardem and best known for his direction of the film classic *Death of a Cyclist*, the movie humorously captured both the desire and enthusiasm of the Spanish people to be "developed" by "Yankee" dollars while the promised capital remained of course glued to the hands of the elites.

¡Bienvenido, Mr Marshall!, carefully navigating around the censor's scissors (which it did not fully succeed in doing), tells the story of the fictional village Villar del Río,[1] which stands for a classic Castilian peasant village. It is dry and dusty; the well on the market square functions only intermittently; the clock tower is broken and time stands frozen at three p.m.; maids gossip, men sit around and chat in the café, women take care of the home or attend church. Peasants toil the land in a sun-drenched landscape for meager rewards. The cunning, hearing-impaired mayor-peasant runs the place in age-old fashion: well-meaning but without a penny to spend.

Figure 6.1
Movie poster for ¡Bienvenido, Mister Marshall! (1953).

An impoverished nobleman, who lives in the shade of his imperial ancestors' glorious past, waits patiently for letters that never arrive.

The pastoral everyday routine of the village is only disturbed by the pending performance in the local café of Carmen Vargas, the "great" budding Andalusian flamenco diva. Her arrival coincides with the unannounced visit of a party of state dignitaries, dressed in black suits and performing the official walk that only government bureaucrats manage to perfect. They announce the planned passage through the village of a high-level U.S. delegation. The functionaries impress on the perplexed mayor that a rapturous welcome of the Americans would be essential for upholding the inviting image Spain wishes to portray and would be of importance to the villagers too. Perhaps even the train line would be extended, not to mention the gifts the villagers would receive from the grateful visiting guests.

The pending visit of the Americans becomes the talk of the village. The schoolteacher lectures to the assembled villagers on the geography and other facts of great American life. The nobleman mutters anti-American sentiments as he reminisces about the imperial defeat of the past century. But they are mostly concerned about how to showcase the village in the best possible light. After considering and dismissing restoring the well and lightening up the waters of the village fountain in festive colors and other schemes to embellish the village, the community decides to dress up their homes and streets in "classic" Andalusian style to correspond to the Americans' image of an authentic traditional Spanish village. At the same time, the villagers draw up a list of desirable things they want from the generous American benefactors. Some would prefer a new bicycle or a sewing machine, others a clarinet or a cow. Only one wish per person can be granted—so general wisdom goes—and many a villager spends anxious nights pondering whether to demand a pair of mules or a new suit to attract the attention of the village beauty at the next ball. The villagers plan a parade for the big day and the mayor nervously prepares his speech in which he will praise the considerable achievements and qualities of Villar del Río. A man is even put in the clock tower to move the timepiece, as a broken clock would surely convey the wrong image to the important guests.

When the great day arrives, the village folks are gathered in their best outfits, the marching band is ready to play, while the mayor nervously wipes the sweat from his brow. As the watchman in the clock tower announces the imminent arrival of the U.S. delegation, a cavalcade of flashy black sedans appears on the horizon, and nerves tense while everyone readies to play their parts. However, the convoy enters the village and, without even slowing down, speeds by "in a cloud of dust, without stopping, leaving the village and its peasants as poor as ever" (Gallo 1974, 224). While the movie captures the mood and conditions of the time well in its painting of poor village life, anticipating the inflow of U.S. dollars and foreshadowing metaphorically the trail these dollars would walk, the movie is also a marked indicator of an imminent radical change in Spain's geopolitical networking.

Indeed, the early to mid-1950s signal a watershed in the structuring of the georelational networks that nurtured and sustained the authoritarian edifice of Franco's Spain. Over the thirty-six years of Franco's rule, the number of dams grew from about 180 in 1939 to over 800 in 1975, reservoir capacity expanded exponentially (see figure 1.3), and the backbone for a nationally integrated system for inter-river basin transfers, which would permit considering the hydro-social cycle as an integrated unitary national cycle (Hernández 1994, 15), was under construction (the Tajo-Segura transfer) by the time Franco died. During the first period, between 1939 and 1955, dam construction was excruciatingly slow and the expansion of irrigation way behind schedule. While 106 new dams were built between 1941 and 1955 (see table 5.1), the capacity of reservoir water only rose from about 4,000 to 8,000 hm^3.

After 1955, there seemed to be no limit on the "progress" made to produce a radically different Spanish waterscape, one that would find its ultimate realization in the construction of the first large inter-river basin transfer designed to balance the uneven hydraulic geography of Spain. Indeed, between 1955 and 1970, a series of mega-dams were built that massively increased the regulatory, hydroelectrical, and irrigation capacity of Spain. A total of 276 dams were completed during this period, while reservoir capacity skyrocketed to 37,000 hm^3 by 1970 and 42,000 hm^3 by 1980. In his speech to commemorate the twentieth anniversary of "Our Movement" and "The Victory," Franco himself insisted how his "great hydraulic and irrigation works are changing the geography of Spain" (Franco 1959c, 1). The changing scalar power geometry through which this transformation was choreographed—one that combined national integration with a new international outlook—will be the leitmotiv of this chapter. It is precisely the insertion of Spain within a new geopolitical imaginary and the engagement of U.S. geopolitical and economic interests within the assemblages of power that sustained the national fascist project that fostered and consolidated Spain's post-1955 hydro-modernization drive.

The Making of a New Spanish Hydraulic World

In October 1937, before Franco had even consolidated his power, he instructed engineer Alfonso Peña Boeuf, who would become minister of public works in 1939, to prepare a *Plan General de Obras Públicas* (General Plan for Public Works), a large part of which would be dedicated to hydraulic infrastructures. Minister Peña Boeuf's proposals were officially approved in 1941 and would provide the framework for hydraulic development during the subsequent decades (Peña Boeuf 1955, 615). This proposed framework basically reiterated the outline of the National Hydrological Plan of 1933, but reoriented its strategy much more decisively to a nationally programmed system aimed at guaranteeing Spain's self-reliant or autarchic development. The plan also was accompanied, most importantly, with the approval of a new law of expropriation for public utility reasons. Land could be expropriated and "occupied" within a few days if the Council of Ministers declared the project in question to be "of national urgency" (Peña Boeuf 1946, 359–360). This removed one of the great historical obstacles to swift state-led implementation of large infrastructural projects.

However, the disastrous financial and political and economic conditions of autarchy during the early Franco reign prevented the massive sociohydrological revolution that had been envisaged in the plan and that continued to be presented as the salvation of the fatherland. There was not enough steel, concrete, money, and machines available to make the waters flow uphill (Peña Boeuf 1946). The only thing that was not in short supply was a cheap, docile, defeated, and impoverished working class.

As discussed in chapter 4, five river basin authorities had already been established during the dictatorship of General Primo de Rivera, one more during the Republic, and the remaining four between 1948 and 1961. However, after the fascist victory, the "regionalist," basin-based perspective that had inspired their foundation in the 1920s was quickly replaced by a nationalist territorial vision. Indeed, the political significance of the "regional" scale became marginalized. Originally intended to be locally organized, regionally focused, participatory, and collective, the existing river basin authorities were abolished in 1942 and replaced by a technocratic-bureaucratic organization in charge of implementing projects that had been planned at the national level in a top-down manner (Alcarez Calvo 1994; Rodríguez Ferrero 2001). The confederations became a "mere technical appendage" of the central Dirección General de Obras Hidráulicas (DGOH) (Pérez Picazo 1999), financed and controlled by the national state. The final blow came in 1959 with the establishment of the water commissioners, who would be in charge of the very powerful administration for assigning and managing water concessions, their policing, and the allocation of irrigation water (Palancar Penella 1960). The preamble to the law explicitly stated the objective of placing the water commissioners under national state control (Melgarejo Moreno 2000, 292–294). The basin authorities literally became the managerial-executive and engineering branch of the Ministry of Public Works. Their remit was limited to taking charge of the technical implementation of plans and works deemed necessary by the national state. The "political" part of the basin authorities was annulled and replaced by a system of technocratic expertise—that of professional engineers, which focused exclusively on "getting things done," on breaking rocks, forging steel, and pouring concrete. This "nationalization" by the state reinforced already strong nationalist and autarchic aspirations to secure Spain's economic self-sufficiency and self-reliant development. "Hydraulic Politics" became "a sublimated expression of the political economy of the nation" (Frutos Mejias 1995, 185) that turned the hydraulic future of Spain into a "national, patriotic, and trans-political mission" (del Moral Ituarte 1996, 181). The DGOH became a well-funded and extraordinarily powerful state department, highly corporatist in ideology and closely associated with key national economic sectors such as engineering offices, construction companies, cement factories, electricity companies, and more (Martínez Gil 1999, 107). The revolving door assured a seamless movement of leading engineers migrating from state administrations to the private sector and vice versa, ever more deeply entangling state and capital accumulation through large-scale public works.

The "concrete and steel fever" and the obsession to leave not a single river "in freedom" shaped the actions and strategies of the DGOH, driven by the totalizing vision of the need to engineer the whole of the nation's river basins as a single, integrated, unified, and national territorial system. It would become a thoroughly corporatist body, the bearers and guardians of a national hydraulic ideology (Martínez Gil 1999). The river basin authorities lost their judicial autonomy, their representational organization, and their integrated planning function. The "jumping of scale" of the institutional and political powers of water management from the river basin to the national scale reinforced a national geographical perspective at the expense of the regional scale.

A great dictator, a voluntarist ideology, engineering plans, corporate support, God's will, and a desire to transform the country's hydraulic structure remain of course insufficient in the absence of concrete, steel, machinery, capital, and specialist know-how if the flow of water is going to be combined with techno-structures that will achieve the restoration of a great Spain. The early Franco era (up to the mid-1950s) was economically one of relative paralysis, enduring shortages, untold misery for many, and sluggish growth. Up to the mid-1950s, very little of what was promised rhetorically was actually implemented, but this soon changed dramatically as the reengineering of Spain's water flows really took off at an accelerating pace. This shift coincided with a profound scalar re-networking of the political-economic networks on which Franco's stability rested. Indeed, Spain's hydro-social modernization and the production of the techno-natural material infrastructures of this modernizing program was predicated upon rescaling the "networks of interest" on which Franco's power rested from a nationalist and autarchic visionary to a more liberal geoeconomic political economy and an internationalist geopolitical imagination, articulated through Spain's integration in the U.S.-led Western Alliance that emerged during the Cold War politics of the second half of the twentieth century. It is this profound rescaling of Spain's networks of interest that we shall turn to next.

Autarchy, National Development, and Geographical Integration: Blood, Sweat, and Tears

The early Francoist period was marked by serious shortages of food and productive equipment, slow growth, and general socioeconomic malaise. The state invariably blamed the "persistent drought" and the isolationist policies imposed by the "rogue democracies" (the United States, the United Kingdom, and France) for the problems. Indeed, the crusade against the *pertinaz sequía* was on par with the struggle against "the judeo-masonic conspiracy" and the "Marxist hordes." Of course, the former was related to serious physical conditions as energy was in dramatically short supply and urban and agricultural water shortages were chronic, primarily as a result of scarce resources to construct new hydraulic infrastructures (Lafuente 2002, 77). In 1949, wages in industry were only 66 percent and in agriculture only 53 percent of those in 1939. Gross domestic product reached only 79 percent of what it had been in 1929 (Cazorla 2000, 27, 268). The value of the peseta had fallen by 50 percent between 1936 and 1949; external debt was mounting and domestic infrastructure seriously damaged.

The political-economic vision of the elites was one that centered on self-generated national autarchic development, achieved through the mobilization of national resources and import substitution industrialization. The enrollment of key elite groups and the rhetoric of nationally integrated development became incorporated in what Raymond Carr called "the permanent ideal of autarchy" (Carr 2001, 156). A strictly controlled market, the freezing of wages, the control of the labor force through its forged integration in the Falangist union, an imposed international isolation because of Franco's wartime support for the Axis powers that turned into an ideology of self-reliant development, characterized the autarchic vision whereby international trade restrictions and a focus on domestic growth were seen as the way to restore Spain's lost grandeur. In the context of this autarchic vision of development, the demand for rapidly increasing irrigation to improve agricultural production was considered to be absolutely vital (Escartin Hernández, Cabezas Calvo-Rubio, and Estrada Lorenzo 1999, 80). Rafael Cavestany y de Anduaga, minister of agriculture, promised in 1951: "irrigated areas have to be extended to the possible limit. We have to search for all possible waters with all means possible" (Cavestany y de Anduaga 1958, 23).

The precarious supply of energy equally pointed to mobilizing Spain's considerable hydropower potential as the basis for its energy independence (Sudriá 1997). This political-economic doctrine of pursuing an independent, nationalist, and autarchic path of economic development had already been suggested by Lorenzo Pardo, head of the National Council for Public Works after the Civil War: "all our contractors are Spanish, Spanish too are the materials, the capital, the employees, the workers, the infrastructure, the tools" (Lorenzo Pardo 1930, 21). Indeed, the views of the hydrotechnocrats and the desires of the authoritarian state fused together almost seamlessly. The making of a nationally integrated economy, so the doctrinal argument went, should be based on mobilizing, almost exclusively, national resources and national capacities (MAPA 1990, 99). Expanding food production through irrigating drylands, combined with supporting hydroelectrical development, became a key priority. The *Plan General de Obras Públicas* of 1939 foresaw the construction of 155 new dams and envisaged an expansion of irrigated lands with 1.25 million hectares (Barciela López and López Ortiz 2003, 65). However, absence of materials, energy, equipment, and, above all, capital made progress in constructing the desired landscape excruciatingly slow. Electricity cuts were rampant until the mid-1950s as hydroelectrical power generation did not expand fast enough, irrigation and colonization progressed slowly (Barciela López and López Ortiz 2000, 362), dam construction was far below expectations, food was rationed, and peasants became even poorer and flocked either to the cities (in particular to Madrid and Barcelona) or emigrated to Northern Europe to work as "gastarbeiter" (guest worker) in the Fordist heartlands of the golden postwar years (Reher 2003). Average income per capita fell from index 100 in 1935 to 82 in 1950 (Gallo 1974, 192–193).

The only commodity not in short supply was labor of a variety of kinds. Indeed, the remaking of the socio-hydrological landscape of Spain is of course one drenched in blood, sweat, and tears. Salaries were only a fraction of what they were before the war and any kind of protest was quickly smothered by ruthless repression. By 1960, the nine Spanish large dam construction companies (see table 5.2) jointly employed sixty thousand workers. On the basis of the captive state-market for large hydraulic infrastructures, some of these companies would go on to become leading global engineering companies.

Moreover, there were tens of thousands of political prisoners: socialists, communists, anarchists, and assorted other undesirables were, if not killed or exiled, held in concentration camps and forcibly put to work, primarily in public works (Molinero, Sala, and Sobrequés 2003). An estimated 250 thousand political prisoners were held in various camps and prisons, often forced to labor on major infrastructure projects. As the law on political prisoners stated, "It is very just that the prisoners contribute with their labor to the reparation of the damages to which they contributed through their collaboration with the Marxist rebellion." For example, for the construction of the *Canal del Bajo Guadalquivir* in Andalucia, over 2,520 political prisoners were mobilized between 1940 and 1962 (Acosta Bono et al. 2004), very often under truly inhumane conditions. Others estimated that more than ten thousand political prisoners were forcibly put to work to dig the canal (Moreno Gómez 2008, 17). Today, this canal is referred to as the *Canal de los Presos* (Canal of the Prisoners) and a monument has been erected to commemorate "Franco's slaves" who dug the canal to permit the irrigation of the land of the Andalusian *latifundistas* (Gutiérrez Molina 2004, 40; 2006) (see figure 6.2). The Guadalquivir River Basin Authority under whose supervision the canal was built could count on both the prison system and the support of the large landowners in the basin to convert their land to highly productive irrigated agriculture, a telltale example of how public effort was mobilized for private gain. The windfall profit did not only result from more intense and competitive agricultural production, but also from the significant increase in the value of the land itself. Today, the canal irrigates almost 80,000 hectares. In the context of an oral history project aimed at unearthing the fate of the prisoners, in her eighties Francisca Adame, whose

Figure 6.2
The *Canal de los Presos* (Canal of the Prisoners) today.

father and brother were forced canal diggers, and who learned to read and write in her seventies, penned a poem to commemorate the Canal of the Prisoners and those who built it (Acosta Bono et al. 2004, 293):

What crime did they commit?
Only they desired equality
of men and peoples.
With the tip of hoe and spade
They made this Canal
quietly and silently,
behind was the warder.
Canal del Bajo Guadalquivir:
That they remove it please.
It is the Canal of the Prisoners,
They made it with their sweat.
. . .
This is not a poem
It is an offering of honor
for all those which were
in concentration camps.[2]

Isaías Lafuente documents how political prisoners were mobilized during the early years of the Franco regime for the construction of major dams (like the dams of the Ebro, Entrepeñas, Pálmaces, Mediano, Riosequillo, Revenga, Barasona, Mansilla de la Sierra, González Lacasa, en El Cenajo) in addition to canalization and irrigation projects (Lafuente 2002). Prisoners were not only used by the state, but also were put at the disposition of large farms and public works companies. Some of the great construction companies established during the Franco period, like Dragados y Construcciones, would also use political prisoners for more than two decades (Soler 2002). Table 6.1 summarizes the available information (which is only now gradually emerging) on the mobilization of political prisoners in the realization of Franco's hydro-political project for Spain.

However, cheap and docile workers and forced labor were clearly not enough to achieve the hydro-modernizing aspirations. The autarchic model did not generate enough capital and equipment to move the earth, forge steel, mix the concrete, and engage the powers of Spain's waters. For that, Spain's leaders had to turn elsewhere and rearrange the coordinates of the nation's geopolitical spatial imagination, its networks of interests, and its geoscalar articulation. The inward-looking nationalist project would have to engage again with the outside world. Of course, Fascist Spain would find a welcoming Western world as soon as the geopolitical realities of the second half of the twentieth century manifested themselves. The remaking of Spain's hydro-social network would have to wait until a repositioning of the geopolitical relations and their associated political economic networking and flows of capital, expertise, and steel took a radical turn after 1953. This moment would prove to be a watershed in terms of permitting the realization of *Paco el Rana*'s hydro-vision for Spain.

Reworking the Nation, Rescaling the Networks of Interests: Yankee Dollars and Dams

Indeed, by the early 1950s, the rhetoric of national autarchy, celebrating the great potential of Spain's development on the basis of self-reliant development that would mobilize and fuse the human and nonhuman resources of the country, sounded increasingly hollow as the material and socioeconomic conditions continued to deteriorate and social unrest, despite relentless repression, threatened to destabilize the regime (Barciela López and López Ortiz 2003; Catalan 2003; Miranda Encarnación 2003). Spain's elites began to recognize that opening up new geographical relations and pursuing the geopolitical insertion of Spain into the Western Alliance was

Table 6.1
Documented use of political prisoners for the construction of hydraulic works (Selection) (1936–1962)

Location	Documented period	Number of workers and (year of reference)	Controlling organization	Type of work
Villatoya (Albecete)	1944–	140 (1944)	Cimentatiaciones y Obras	Construction of bridge
San Adrián del Besós	1944–	70 (1944)	Cimentatiociones y Obras	Construction of bridge
Celis (Cantabria)	1949–1950	54 (1949) 138 (1950)	-	Hydroelectrical
Celucos (Cantabria)	1949–	?	Dragados y Construcciones	Hydroelectrical
Pálmaces de Jadraque (Guadalajara)	1942–1946	50 (1942) 50 (1945)	Private: ECIA	Dam construction
Irún (Guipúzcoa)	1944–	134 (1944)	Ferrocarriles y Construcciones ABC	Canal
Barasona (Huesca)	1946–1949	180 (1948)	Cimentaciones y Obras	Dam construction
Guara (Huesca)	1962–	20 (1962)	Cimentaciones y Obras	Dam construction
Madiano (Huesca)	1943–1955	50 (1943) 68 (1954)	Vías y Riego; Dragados	Dam construction
Barrios de Luna (Léon)	1952–1955	40 (1954)	Herederos de Ginés Navarro	Tunnel for dam Hydroelectricity
Buitrago del Lozoya (Madrid)	1944–1952	250 (1945) 123 (1949)	State	Dam construction
Escorial, El (Madrid)	1944	50 (1944)	Private: San Román	Water supply
Patones (Madrid)	1957–1960	62 (1958) 94 (1961)	Construcción AMSA	Water supply

Location	Years	Amount (Year)	Company	Type of work
Cenajo, El (Murcia)	1952–1957	30 (1952) 70 (1955)	Construcciones Civiles, SA	Dam construction
Orense	1952–1953	400 (1952)	Dragados y Construcciones	Dam construction
Reinoso de Cerrato (Palencia)	1944–	50 (1944)	Cimentaciones y Obras	Bridge over river
Anguiano (La Rioja)	n/a	170 (1944)	Construcciones ABC	Dam construction
Mansilla (La Rioja)	1949–1958	65 (1949) 50 (1955)	Ingeniería y Construcciones Marcor, SA	Dam construction
Ortigosa de Cameros (La Rioja)	1953–1962	60 (1956) 555 (1958) 88 (1961)	Ereño y Cia., SA	Dam construction
Arroyo (Santander)	1943–1949	258 (1943)	Vías y Riegos	Dam construction
Revenga (Santander)	1947–1950	100 (1949)	State	Dam construction
Segovia	?	?	?	Dams and irrigation
Puebla del Río (Sevilla)	1952–1955	105 (1953)	State and private	Agricultural transformation
Castillejo (Toledo)	–1954	42 (1954)	Cimentataciones y Obras	Bridge over Tajo
Puerto del Rey (Toledo)	1944–	50 (1944)	Hnos. Nicolás Gómez	Canalization
Talavera de la Reina (Toledo)	1942–	342 (1942)	Hnos. Nicolás Gómez	Dam construction
Chelva (Valencia)	1941–	300 (1941)	Portolés y Cia.	Dam construction
Valladolid	?	?		Dam and irrigation canals
Rentería (Vizcaya)	1944–	135 (1944)	Construcciones ABC	Canalization
Freson de la Rivera (Zamora)	1945–1946	95 (1945)	Don Ramón Echave	Irrigation
Tauste (Zaragoza)	1956–1959	11 (1959)	Bernal Pareja SA	Irrigation

Source: Based on Acosta Bono et al. 2004, 65–75.

vital in order to secure not only the modernization of Spain, but also the longevity of the dictatorial regime. Strategically extending the spatial reach of the networks of interests on which the regime rested was pivotal to pursue the modernizing project envisaged for Spain.

The Spanish ruling elites understood the realities of the emerging geopolitical order choreographed by Cold War strategists and eyed the United States, whose gaze also started to turn to Spain as a possible ally in the new geopolitical strategic geometry of the postwar order. The articulation of national elite interests with an internationalizing agenda was a tricky affair. Nazi Germany had supported Franco militarily during the Civil War. The killer aerial raid by the German Condor Legion and the Italian Fascist Aviazione Legionaria, possibly the first bombing of a civilian population, had obliterated the Basque town of Guernica (later immortalized in Pablo Picasso's painting of the same name). While Franco tried to ally himself directly with Hitler during the war, the latter did not show much enthusiasm given Spain's weak military and economic capacity. Nonetheless, Franco actively supported the Axis powers. The *División Azul* (Blue Division), a volunteer legion of over eighteen thousand soldiers was sent to fight with the Nazis on the Eastern Front. After the war, Franco tried to distance himself somewhat from that history, insisting on the neutral stance Spain took during the war. This wartime legacy and the nationalist rhetoric of early Franquism were accompanied by an isolation of Spain from the newly established international organizations and exclusion from the postwar Western-allied networks. In 1940, the United States had already embargoed the export of oil to Spain. In 1945, Franklin D. Roosevelt confirmed Spain's pariah status in a letter to the American ambassador in Spain:

Having been helped to power by Fascist Italy and Nazi Germany, and having patterned itself along totalitarian lines, the present regime in Spain is naturally the subject of distrust by a great many American citizens who find it difficult to see the justification for this country to continue to maintain relations with such a regime. . . . Most certainly we do not forget Spain's official position with and assistance to our Axis enemies at a time when the fortunes of war were less favourable to us. . . . The fact that our Government maintains formal diplomatic relations with the present Spanish regime should not be interpreted by anyone to imply approval of that regime and its sole party, the Falange, which has been openly hostile to the United States and which has tried to spread its fascist party ideas in the Western Hemisphere. Our victory over Germany will carry with it the extermination of Nazi and similar ideologies. . . . I can see no place in the community of nations for governments founded on fascist principles.[3]

In March 1946, the United States, France, and Britain jointly condemned the Franco regime (Liedtke 1998, 17). In June 1946, the United Nations Security Council subcommittee on "The Spanish Question" evidenced conclusively Franco's pro-Axis support (United Nations Security Council 1946). A few months later, the General Assembly excluded Spain from all UN-related organizations and recommended all its members to withdraw its ambassadors from Spain.

However, the tide would begin to turn quickly as the United States' anti-communist stance combined with its newfound role as the Western world's geopolitical arbiter opened up new possibilities for enhancing Spain's international respectability. In Spain, nationalist rhetoric too had to be aligned with opening up to the once-dreaded "vile" democracies of the West, while U.S. foreign policy needed to deal with the diplomatic and ideological difficulties of diverting from the early postwar isolationist consensus on Spain. A slow discursive shift could be discerned, both in Spain and abroad, that began to define the Spanish regime no longer as "fascist" but as authoritarian, a *Caudillismo*, characterized by strong leadership in a one-party state, but one that had only superficial similarities with fascism. The latter became primarily identified with Hitler's Nazi regime and the extermination of Jews.

Defusing foreign hostility domestically had to go hand in glove with satisfying the interests of the three key pillars on which the regime's stability rested: the powerful and internationally influential Roman Catholic Church, the military, and the economic elites. The first was achieved already in 1953 when Franco signed a concordat (the last of its kind) with Pope Pius XII that granted the Spanish Catholic Church extraordinary and exclusive privileges, such as exemption from government taxation, subsidies for new building construction, censorship of materials the church deemed offensive, the right to establish universities, the right to operate radio stations and to publish newspapers and magazines, protection from police intrusion into church properties, and exemption of the clergy from military service (NN 1953). This restored the prestige of Spain among the very powerful international Catholic movement.

Opening up to the United States, which was in turn seeking to extend and solidify the new cartography of the Western Alliance, allowed Spain to circumvent the continuous hostility of the other European powers and to perforate the isolationism of Spain. Moreover, improving relationships with the United States could simultaneously serve a military concerned with strengthening the defensive capabilities of Spain while obtaining much needed financial assistance for the modernization of its military equipment (Termis Soto 2005). Between 1950 and 1953, the sedimentation of the Cold War as the axis around which postwar geopolitics was choreographed permitted indeed a rapid rapprochement between the United States and Spain. The United States chose symptomatically to forget its earlier enmity toward fascist Spain and increasingly began to play the role of Spain's ambassador in international forums against the official policies of, among others, France and the UK, which opposed the integration of fascist Spain in the newly established postwar international political organizations.[4] From 1947 onward, the United States slowly moved, under the Truman administration, toward "a pragmatic policy of accommodation" (Rosendorf 2006, 375).

As early as 1948, Spain had sent José Felix Lequerica—who the United States had rejected in 1945 as Spain's new ambassador because he was considered an "unabashed fascist," "being more German than the Germans"—officially as the Spanish foreign service inspector, but really to organize Spain's lobby efforts in the United States. He immediately hired the services of the law firm of Cummings, Stanley, Truitt and Cross to represent the interests of Spain. This was a well-considered choice of lobbying firm. Truitt was the in-law of Senator and soon to be Vice President Alben Barkley, while Homer Cummings had served for six years as Roosevelt's attorney general. In addition, Lequerica retained the services of Charles Patrick Clark as his main lobbyist for the retainer of $100,000 a year. In a vitriolic attack columnist Drew Pearson stated that Clark "has helped wangle money [from Congress] at a faster rate than Franco has been able to spend it" (Pearson 1952, 6).

The Spanish lobby in the United States became indeed a highly influential body, consisting of a powerful bipartisan group of Americans that brought together Catholics (such as Jesuit and notorious Franco supporter Joseph F. Thorning (see Thorning [1943] 1968), wheat and cotton producers, anti-communists, military strategists, and influential business leaders who sought closer ties with Spain. The Archbishop of New York, Francis Cardinal Spellman, U.S. Representatives Alvin O'Konski and Eugene Keogh, U.S. Senators Pat McCarran, Chan Gurney, and Joe McCarthy, as well as James Farley, Coca-Cola executive and high-ranking Democrat, were among the many political, banking, steel, timber, and cotton business leaders joining the ranks of those advocating closer ties with Spain. They formed a formidable and effective lobby (see, among others, Viñas 1981; Areilza de 1984; Payne 1987; Briggs 1994; Byrnes 1999; Termis Soto 2005).

While the prospects for Spain's international rehabilitation seemed more remote than ever with the surprise 1948 presidential victory of Harry

Truman, who had declared that "Franco was as bad a dictator as Mussolini or Hitler," pro-Spanish lobbying continued unabated (Byrnes 1999, 266; Preston 1995, 597). Indeed, while Truman still insisted, on November 2, 1950, that "it would be a long, long time before there is an ambassador in Spain," the State Department announced just a few weeks later that Stanto Griffis would be appointed to that post. All this happened despite the fact that Franco once said "what joy to see the German bombers one day punishing the insolence of the skyscrapers of New York" (cited in Byrnes 1999, 263). The U.S. Navy had already called on Spanish ports in 1949 and Chief of Naval Operations Forrest P. Sherman visited Franco in 1951 to discuss mutual cooperation. In the meantime, U.S. economic interests had begun to explore Spain's untapped investment possibilities. National City Bank had already granted a $25 million loan to Spain in 1949, and the U.S. government had approved a loan of $62.5 million in 1950 "without demanding in return democratizing efforts from the part of the Franco Regime" (Barciela López et al. 2001, 159). American Express set up offices in Madrid in 1951 and Amex's CEO Ralph Reed met with the Caudillo to cement their collaboration (Rosendorf 2006). Trans World Airlines ("your skyway to gay, romantic Spain") and the Hilton Hotels chain would soon follow suit. Conrad Hilton honored Franco's efforts by establishing his first European Hilton Hotel in Madrid and clearly saw his efforts as part of the wider Cold War priority of keeping communism at bay (Rosendorf 2006). When the hotel opened in 1953, he offered, in immaculate Spanish, his unabashed support for Franco's project:

The Western World owes a debt of gratitude to Spain and her people for many things over the centuries. But in my mind, in this lovely summer of 1953, she stands on a glorious pedestal of the twentieth century for being the only nation in the world which has defeated Communism. Russia was already swallowed up by this monstrous thing, so was Czechoslovakia, Hungary and Poland . . . the pincers were to close over all of Europe. The world should be tremendously grateful to Spain for the great sacrifice she made in hitting back so hard that the Communist time-table has been upset for ever since. (cited in Bolton 1954, 123; see also Rosendorf 2006, 390)

Soon thereafter, millions of tourists would start to flock to Spain, laying part of the foundations for Spain's 1960s economic miracle. In 1952 alone, foreign tourism revenue already amounted to 100 million pesetas in foreign exchange (Rosendorf 2006), roughly equivalent to $41 million. By 1960, it would be close to $300 million, and by the time of Franco's death it had reached $3.5 billion (Barciela López et al. 2001, 199, 447; Figuerola Paloma 1999). Of course, these tourists all required being able to bathe, wash, and drink—something not evident in cities like Madrid where water service was suspended 53 percent of the time in 1949 (March Corbella 2010, 481). The tourist infrastructure development in the Levant also required huge quantities of water, some of which would have to be imported from elsewhere (discussion follows). The urbanization of water would indeed become a key priority during the Franco era despite the recurrent rhetoric of the need to expand irrigation water sources.

In November 1950, the United Nations boycott was repealed and Spain scored its first diplomatic victory as a full member of the international community on its admission in 1952 to UNESCO, the United Nations Educational, Scientific and Cultural Organization. Spain entered the United Nations in December 1955, joined the International Financial Corporation in March 1960 (*Diario ABC*, March 25, 1960), and became a full member of the World Bank and the International Monetary Fund in 1958. The most significant moment was undoubtedly the signing of the secret "Pact of Madrid" in September 1953 by Alberto Martín Artajo, Spanish minister of foreign affairs and James Dunn, U.S. ambassador, in which Spain agreed to let the United States use parts of Spain's territory for military bases in exchange for economic, military, and technical aid[5] (Viñas 1981; Guirao 1998; Liedtke 1998). U.S. financial support and investment was about to flow into Spain. Together with the diplomatic and economic relations that came with the implementation of the aid program, these pacts would provide the (financial) bedrock for the years of rapid growth and modernization of the later 1950s and 1960s (Niño 2003). Indeed, while primary materials and industrial equipment were extremely scarce until the mid-1950s, the inflow of U.S. aid permitted the rapid development of infrastructure. Quite literally, the presence of U.S. military bases in Spain was bought in exchange for dollars and equipment that would, among other things, realize Franco's techno-natural project. Dam construction skyrocketed to new heights after the mid-1950s. The international scalar networking in which Spain inserted itself in turn contributed to the hydraulic geographical integration that the national blueprints of Franco and his engineers had envisaged. The Economic Assistance Agreement with the United States also required Spain to embark on an economic liberalization trajectory: it stipulated that Spain had to "stabilize its currency, establish or maintain valid exchange rates, balance its government budget . . . and encourage competition" (U.S. Department of State 1953, 437).

The deal with the United States took a form similar to the Marshall Plan (from which Spain was excluded because of its wartime support for the Axis powers). The three executive agreements of 1953 covered defense, economic cooperation, and technical assistance. The financial support of the United States was earmarked as follows: 10 percent for administrative expenses, 60 percent for the military bases, and 30 percent for financial aid. From 1958 onward, 90 percent of the funds would take the form of financial aid (Suárez Fernández 1984, 72–73). During the first period $465 million of assistance was provided ($350 million for military help and $115 million for economic and technical assistance [Termis Soto 2005, 48]). The agreement of course opened the door wide for Spain's international recognition. Gradually, the fascist-nationalist tone was massaged and replaced by an increasingly ambiguous discourse and course of action that maintained a rigorously nationalist-conservative tone at home, combined with sustained repression of dissident voices, but a pragmatic and more accommodating perspective toward dealing with the outside world. This became more deeply entrenched as a generation of conservative Catholic Opus Dei–linked technocrats who embodied precisely this mixture of Catholic conservatism, staunch anti-communism, and international-liberal views on economic policies were appointed to key governmental positions in 1957. They expressed the desire of the national economic elites and "modernizing" state officials to expand the scalar reach of Franco's "networks of interest." As the Minister of Commerce Alberto Ullastres wrote in 1960, "Spain must not be left in the margin of these integrating movements . . . otherwise the nation would more and more become a backward province in Europe" (cited in Baklanoff 1976, 751).

The liberal, yet interventionist, Stabilization Plan of 1959 further consolidated the reworking of Spain's geopolitical and geoeconomic relations. The international assistance for Spain granted in the aftermath of the acceptance of the plan amounted to over $540 million (Sardà 1970, 473). Spain's admission to the Organisation for European Economic Co-operation in Europe (later the Organisation for Economic Co-operation and Development [OECD]) on July 21, 1959 was greeted with triumphant reports, announcing that the Spanish government would receive $490 million of international aid (*Diario ABC*, July 21, 1959). Just a few days earlier, the media reported how "an extraordinary homage was given to Spain and to Generalissimo Franco in the U.S. Congress by Senator Mansfield and a large number of representatives, among whom Montoya, Anfuso, Kegon, and Mutter." By referring to the anniversary of the Spanish uprising and Fascist victory of 1939, they confirmed that "the National Movement constituted a decisive contribution for achieving an anti-communist Europe and avoiding that a large part of the West fell into the Soviet zone of influence" (*Diario ABC*, July 18, 1959, 35). Democratic representative Victor Anfuso reportedly put it as follows: "If it were not for the efforts made over twenty-three years, Spain would be communist today . . . in that respect we are indebted to the Spanish people: a debt of gratitude. Today, we all know that Spain is the greatest friend and ally of the United States. . . . There is no reason therefore to deny Spain her earned right to full participation in the Western defense systems."[6]

In the same session, Styles Bridges, while celebrating the virtues of Franco, engaged in a wonderful example of revisionist history when he urged "not to forget how Generalissimo Franco maintained the neutrality of his country during the five years of the Second World War" (*Diario ABC*, July 18, 1959, 35). Just a few months later, Franco would nonetheless reiterate his disdain for democracy and reassert the great advances the National Falangist Movement had made (*Diario ABC*, October 30, 1959).

In sum, extending the spatial reach of the networks of interests, combined with a liberalizing economic course, not only reaffirmed Franco's position, but also permitted new flows of money, materials, and expertise to merge with water in the construction of a radically revolutionized hydro-social edifice. Between 1951 and 1963, a generous total of more than $1.3 billion was granted to Spain in the form of economic aid (Calvo 1998, 2001). U.S. assistance provided Franco with political support and kudos to muscle his way into the Western Atlantic alliances and networks, permitted him to modernize the country militarily, opened up the economy at the expense of the advocates (among them Franco himself) of self-sufficiency, and consolidated the regime while demoralizing and further marginalizing both internal opposition and the international anti-fascist movements (Niño 2003, 26–27). For the United States, the economic stabilization of Spain would further entrench the power of Franco and ensure the continuing anti-communist stance of Spain, erode the danger of a left-wing insurgency, and secure a vital territorial military presence against the Eastern Bloc. The international flows of capital that arrived in Spain between 1957 and 1973 in addition to the official aid were extraordinary: $18.5 billion from tourist receipts, $7 billion from emigrant remittances, and an estimated net foreign capital investment of almost $6 billion (Baklanoff 1976, 754).

A large part of the financial aid went to agricultural machinery, steel, electrical equipment, and infrastructure. Of the Spanish counterpart funding, most investment was directed toward agricultural irrigation projects, railroads, and hydraulic works (Fernández de Valderrama 1964; Puig 2003, 114). Of the direct financial aid, almost a quarter was invested in capital goods and infrastructure. Through this, Americans and Spaniards "wove and strengthened social networks with local, national and international reach" (Puig 2003, 117). The scalar rearrangement of the networks the dictatorship wrought proved vital in the process of autocratic modernization (Puig and Alvaro 2002).

The scalar extension of Spain's political and financial networks, in turn, facilitated and nurtured the hydro-social transformation of Spain's physical and socioeconomic geography, a project pursued by all possible means until the end of the Franco regime (and beyond). There seemed to be no limit on the "progress" made to produce a radically different Spanish waterscape, one that would find its ultimate realization in the construction of the first large inter-river basin transfer scheme, the backbone for producing a national water system and for balancing finally the uneven hydraulic geography of Spain. The great leap forward happened after 1955, with 276 dams built between then and 1970. During the same period, the total volume of water reservoirs capacity skyrocketed exponentially from 8.3 billion cubic meters to 36.9 billion, to reach 42 billion by 1980 (Ministerio de Agricultura 1980; Birch, Levidow, and Papaioannou 2010). Indeed, between 1955 and 1970, a series of mega-dams were built that massively increased the regulatory, hydroelectrical, and irrigation capacity of Spain. For example, in 1960 the dam complex of Entrepeñas/Buendia, with a total reservoir capacity greater than the combined national capacity of 1944, was inaugurated on the head flow of the Tajo River. This complex would later become the starting point for the first big inter-river basin transfer scheme (discussion follows). A high-level visit of the commissioner of the U.S. Bureau of Reclamation and several of his aides to Spain in 1964 resulted in a major publication that lauded Spain's hydraulic achievements: "Spain stands in fifth place among the nations of the world in number of large dams and in third place in the number of dams completed in the past 25 years or now under construction, following only the United States and Japan in this regard" (cited in del Campo y Francés 1992, 179).

In his speech to commemorate the twentieth anniversary of "Our Movement" and "The Victory," Franco himself insisted how his "great hydraulic and irrigation works are changing the geography of Spain" (Franco 1959c, 1). This modernizing project is only possible because "nature concentrated the possibility of those great hydraulic and irrigation works which today are changing Spain's geography.... They have provided the foundation for this reconstruction of Spain, for this aggrandizement of the Fatherland"[7] (*Diario ABC*, November 4, 1959, 1). Still insisting on the "national" capacities of Spain, he conveniently forgot to mention the vital role of the changing global geopolitical configuration. One of the most emblematic projects of overhauling Spain's water landscape would be the Tajo-Segura water transfer, the biggest water transfer project ever undertaken in Europe.

The Tajo-Segura Transfer: Epitome of a National Water Imaginary

The pillar for the construction of a nationally integrated hydro-social system would be the first large inter-river basin water transfer scheme: the Tajo-Segura complex. This extraordinary project constitutes both the epitome of twentieth-century Spanish hydro-politics and the material and symbolic point around which much of the debate and hydro-social practices over the next forty years would be articulated. It constitutes the inaugural moment for the construction of an integrated water network that would produce a territorially interconnected national water grid whose main function was to carry water from regions with too much of it to the arid lands of the Levant that have too little (see chapter 7). In the process, of course, it would reinforce and solidify Spain as one and undivided, linking the diverse regions and ecologies of the peninsula in political and hydro-social solidarity and cohesion. As Franco affirmed in his speech delivered for the inauguration of the dams of Cenajo and Camarillas on June 6, 1963 in Murcia, "our politics possesses the great virtues of unity" whereby "the gold of our lands is put in the hands of our men" (*La Verdad*, June 7, 1963, 1).

The basic idea of transferring water from basins with "excess" or "surplus" water to basins with a water "deficit" had already been envisaged by Lorenzo Pardo in the 1933 first National Hydrological Plan (Lorenzo Pardo [1933] 1999) (see chapter 4):

> If the ideas of Joaquin Costa were based on the unity of the river basin as the framework for the implementation of hydraulic projects, Hydrologic Planning [after 1933] extended this framework to the national scale, by advancing as one of its objectives the correction of the existing disequilibria on the Iberian Peninsula by means of interconnecting the river basins.[8] (Melgarejo Moreno 2000, 273)

In the postwar period, the idea of large-scale water transfers was taken up again and became an integral part of the ideological and political mission to unify the fatherland, both materially and spiritually, while maximizing the socioeconomic development potential of all regions through reallocating water "justly" and "evenly." The implementation of the project, however, could not be envisaged without a transformation in the political-economic configuration that would permit mobilizing the necessary capital to carry out such a gargantuan task. From the mid-1950s onward, the conditions were ripe to write the last chapter of *Paco el Rana*'s wet dream for Spain. In a hagiographic account of Franco's great contributions, Suárez Fernández recalled and celebrated the government's decision to change the hydraulic structure of the country and to construct a nationally integrated system of interconnected river basins:

The council of Ministers of 1 June 1955 examined the most ambitious project in the History of Spain: changing its hydraulic structure. Industrial and agricultural prosperity could not be trusted to the seasonal and erratic rains. The constructed dams demonstrate the efficiency of dammed water. Now, it is a question of utilizing the waters of the Tajo river as a great reserve for redistribution, damming the same waters several times over, returning and distributing [the water] to irrigate fields each time further away. . . . On 14 July 1955, Franco inaugurated the great reservoirs of Entrepeñas and Buendía, in the province of Guadalajara. He called them, proudly and significantly, "the sea of Castilla." Therefore, the subsequent phase of hydraulic politics—to be realised over decades and with persistence—consisted precisely in the making of interior seas by means of networks of inter-basin water transfers.[9] (Suárez Fernández 1984, 227–228)

As the director general of public works in a commemorative speech for the foundation of the Centro de Estudios Hidrográficos (1963) in 1971 reiterated, "The National Plan of Hydraulic Works [of 1933] considered the country as single hydrographical unity" (Director General de Obras Hidráulicas 1971, 396). "Over the last ten years," he continued, "we have been able to establish a hydraulic balance and initiated the programs to correct the peninsular hydrographical disequilibrium" (396–397).

Not only Lorenzo Pardo had suggested large-scale water transfers. A host of other plans and proposals, mainly for political propagandistic reasons (Gil Olcina 1992, 19), had circulated during the 1930s, among others from Félix de los Rios, then director of the *Confederación del Ebro* (Ebro River Basin Authority), and from the national delegation of technical services of the Fascist Party (*Falange Española Tradicionalista y de las J.O.N.S.*), which would later become Franco's official national state party (Preston 1995). Both proposed massive transfers of water from the Ebro to the Levant (Torres Martínez 1961) and its implementation would be a recurrent demand of the conservative elites of the Levantine regions where Franco's power was very deeply and unwaveringly entrenched. These proposals would, in a modified form, again surface in the contested National Hydrological Plan of the Partido Popular in 2000 (Ministerio de Medio Ambiente 2000a,b) (see chapter 7).

The political decision to go ahead with the transfer was taken by the council of ministers in 1955, but the actual works did not start until 1968 (López Bermúdez 1974). When, in 1963, Franco inaugurated the Centre for Hydrographical studies (*Centro de Estudios Hidrográficos*) (ROP 1963, 553; Urbistondo 1963), one of its first missions was to undertake preliminary studies for the Tajo-Segura and other possible water transfers. This vision "announced the end of the old concept of the hermetic boundaries of river basins. Water was from now onwards considered to be a national good that has to be taken to where it is most productive and most scarce" (Saenz Garcia 1967, 190). Franco announced publicly the decision to construct the Tajo Segura Transfer at the June 1963 inauguration of the dams of Cenajo and Camarillas in Murcia, the key beneficiary of the transferred water (Pérez Crespo 2009). On July 30, 1966, the government ordered the preparation of a transfer project proposal.

On January 30, 1967, then-Minister of Public Works Federico Silva Muñoz announced in his famous speech in the *Teatro Romeo* in Murcia (and repeated later in Cartagena and Alicante), that the Council of Ministers had approved the project, which would, he argued, rectify the hydraulic imbalance between the different Spanish river basins: "The Southeast fought a hard battle with water, one that extended over the centuries. This, which is yet to be written, is of epic dimensions. The hydraulic operation that we are going to undertake is a great enterprise of national justice"[10] (Silva Muñoz, cited in Pérez Crespo 2009, 50).

The pre-project planning documents were completed in November 1967 with a proposed timeline for the transfer (figure 6.3). The preamble calls it a "General Plan of Correction," which is "the sensible thing to do." The plan opens by stating yet again that the temporal irregularity and unfavorable geographical distribution of water necessitated consideration of the combined and integrated management of the water resources of various river basins that are complementary in terms of demand and supply (Ministerio de Obras Públicas 1967, 1). On February 5, 1968, the project was formally approved, and the Council of Ministers ordered work to begin on September 13 of the same year (González Paz 1970, 987). The aqueduct takes water from the reservoir of the Buendía dam on the Tajo River east of Madrid and is pumped and channeled to the Alarcón reservoir on the headwaters of the Jucar River, north of Albacete; from there it is moved to the headwaters of the Segura River that flows toward Murcia. Water is pumped over a height of 300 meters and flows over a distance of 286 kilometers, of which 69 km is tunneled (32 km traverses the Sierra de Hellín, which separates the Júcar and Segura basins, at a depth of 300 meters), 11 km via an aqueduct, and the remainder through an open-air canal (Gomez De Pablos 1972, 471). The transfer system has a theoretical capacity of 33 m^3 per second, enough to move an anticipated final annual volume of 1,000 hm^3 (see figure 6.4).

On June 19, 1971, a law was passed to permit the transfer of water from one basin to another. In the first phase, 600 hm^3 would be transferred

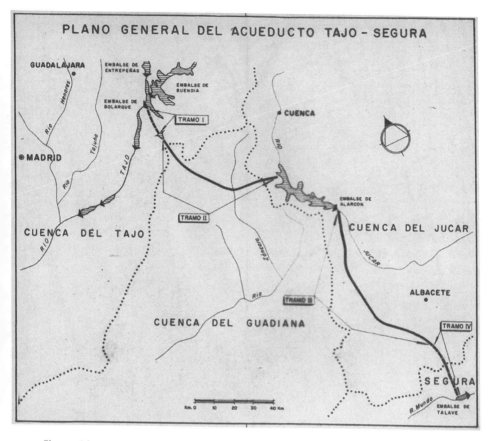

Figure 6.3
General plan of the Tajo-Segura water transfer (original version, 1967).

annually from the Tajo River to the reservoirs of the Segura basin with an eye toward irrigation and urban supply in the provinces of Murcia and Alicante: 400 hm^3 was earmarked for irrigation purposes and 110 hm^3 for urban supply (García Yelo 1997). In the second phase, the project would be upgraded and the transferred volume would increase to 1000 hm^3 per year. However, the latter has never (or has not yet) been implemented. Then Minister of Public Works Gonzalo Fernández de la Mora invoked again the metaphor of "hydraulic surgery" to refer to these "most important works in the hydraulic history of Spain." He continued, "All this we have enjoyed during the most expansive period of our history, thanks to this exceptional man who is Francisco Franco" (Gonzalo Fernández de la Mora 1971, 338–339). After Franco's death, the works continued, albeit at a slower pace.

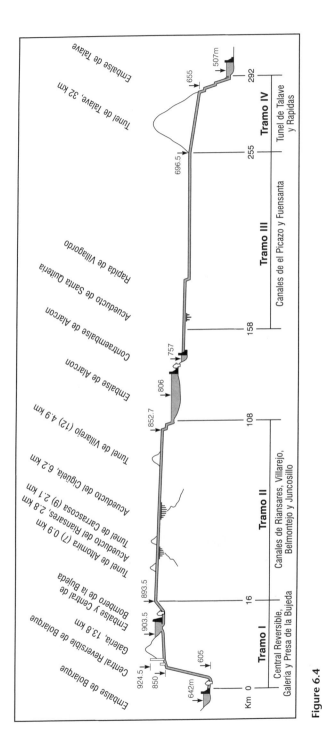

Figure 6.4
Transect and final trajectory of the Tajo-Segura water transfer.
Source: http://www.chsegura.es/chs/ (accessed July 15, 2011).

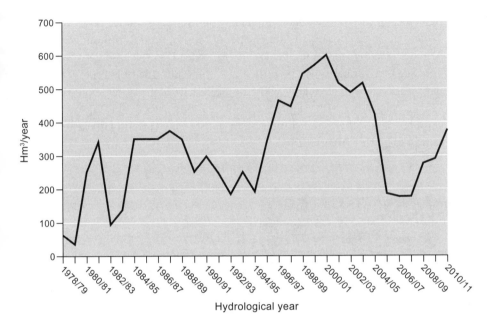

Figure 6.5
Annual water transfers, Tajo-Segura Aqueduct, 1978–2011.
Source: http://www.chsegura.es/chs_en/cuenca/resumendedatosbasicos/recurso shidricos/trasvaseTajoSegura.html (accessed August 21, 2013).

Even after the aqueduct was completed in 1979, rarely did the volume of transferred water reached the legally permitted annual transfer volume of 600 hm^3. As figure 6.5 shows, actual annual water transfers were generally far below the legally permitted maximum.

The elites and irrigators of the Levant, who allied mainly with the newly established UCD party (*Unión de Centro Democrático*, or Union of the Democratic Centre) after Franco's death and later with the conservative *Partido Popular* (Popular Party), pursued relentlessly their lobbying efforts to hasten the completion of the Tajo-Segura complex. The fear of increasing emigration from the region, the need to render agriculture competitive with the European Common Market (of which Spain was still not a part—it would join in 1986), and the relentless promotion of intensive tourism were added to the well-rehearsed arguments in favor of the transfer. However, the demands for regional autonomy, the resistance of the region Castilla-La Mancha to the transfers of its waters to other regions, and the association of the project with "the interests of Franquismo" (*La Linea*, October 12, 1977) stalled the completion and expansion of the transfer. When the final trajectory of the transfer was completed (the tunnel of Talave) in 1978, *La Verdad*, the local conservative newspaper of Murcia, proudly announced "It Arrived!" (*La Verdad*, September 24, 1978). Indeed, so the argument went, "the Segura river basin will be turned into the California of Europe," comparing the transfer favorably with the California State Water Project (*La Verdad*, August 6, 1978). Figure 6.6 shows various parts of the Tajo-Segura complex. Four other, smaller inter-basin transfers are currently in existence as well (del Moral Ituarte 1994).

Figure 6.6
The Tajo-Segura Water Transfer Project. (a) Tajo-Segura transfer, Alcázar del Rey, Cuenca, Spain; (b) Tajo-Segura transfer open channel, Balazote, Albacete; (c) Tajo-Segura transfer, Los Anguijes Reservoir, end of water transfer open section and beginning of Talave tunnel; (d) Tajo-Segura transfer, Sifón de Orihuela.
Sources: (a) http://commons.wikimedia.org/wiki/File:Trasvase_Tajo-Segura.jpg (accessed August 21, 2013). (b) http://commons.wikimedia.org/wiki/File:Trasvase_Tajo-Segura_por_Albacete.jpg (accessed August 21, 2013). (c) http://commons.wikimedia.org/wiki/File:Presa_de_Los_Anguijes.jpg (accessed August 21, 2013. (d) A. Amorós, Diario Información (Alicante).

CODA: The High Price of Hydro-Progress or . . . the Forgotten Tragedy of Franco's Dams

The dramatic transformation of the hydro-social landscape under fascism was marked by all manner of ruthless expropriations, the silencing of oppositional voices, the mobilization of forms of slave labor, and "natural" disasters. The total control of the regime over the media assured that such practices were largely ignored or minimized. One of the most emblematic—but still largely forgotten—moments that showed the brutal face of Franco's hydro-technical project occurred on the freezing minus-18-degree Celsius night of January 9, 1959.[11] At around midnight, the dam of Vega de Tera near the village of Ribadelago in the region Castilla-León broke and eight million cubic meters of water thundered through a 140 meters-wide gap over the 549 sleeping villagers. The dam's construction by Hidroeléctrica Moncabril (later taken over by Unión Fenosa) had started in 1954 and was inaugurated by Franco in September 1956. A total of 1,300 men labored on the dam for a salary of 9 pesetas ($0.15 in 1959 exchange rates) per day (NN. 2005: 7). When the dam's wall collapsed, the village disappeared for fourteen minutes under nine meters of water, taking residents and 75 percent of all livestock into the nearby Lake Sanabria. Most of the village was totally destroyed (see figure 6.7). In the avalanche, 144 people died, more than half of them children. Only twenty-eight corpses were ever recovered and that number was recorded as the official death toll (García Díez 2001). The others are still buried in the debris today; aging parents continue to visit the lake where their children found an untimely grave. NO-DO reported the tragedy briefly in its reel of January 19, 1959. It was the penultimate news item and exactly one minute and thirty-eight seconds of reel were devoted to the tragedy.[12] The print media first reported that the dam had "suddenly overflowed" because of heavy rainfall (*Diario ABC*, January 10, 1959; *La Vanguarda*, January 13, 1959). Only days after the tragedy, the breach of the dam was mentioned, but attention quickly moved to praise the great support and international financial help that flowed to the area to deal with what *Sábado Gráfico*, a popular illustrated magazine, called "the Pompeii of water" (García Lozano 2009). Franco adopted the village the day after the tragedy (*Diario ABC*, January 10, 1959, 19) and a new settlement, Ribadelago de Franco, named in honor of its new godfather, was built a mile from the devastated village. Only about a hundred people moved to the new settlement. Many of the other residents, having lost family and their livelihoods, emigrated. The state paid 95,000 pesetas ($1,583 at 1959 exchange rates) for each deceased man, 80,000 pesetas ($1,333) for each

a

b

c

Figure 6.7
The punctured Vega de Tera dam (a) and ruined village of Ribadelago (b and c), Zamora, 1959.
Source: Junta de Castilla y Léon, Archivos de Castilla y Léon, Fondo Gobierno Civil, Subdelegación del Gobierno, Signatura 1/06, 1/10, and 1/16.

woman and 25,000 pesetas ($416) for each child as compensation for the loss (Lera 1999). As entire families had died or emigrated, much of the reparation money was never paid out. In 1963, the court case concluded that the disaster was the result of a combination of deficient construction and use of inferior materials, differential behavior of steel and granite under extreme low temperatures and excessive rain- and snowfall just prior to the catastrophe. The construction company was ordered to pay 19 million pesetas. The managing director of the company, two engineers and a technician, who were directly responsible for overseeing the construction, were found guilty of "reckless imprudence" and sentenced to one year of imprisonment. They appealed the conviction and were pardoned (NN 2005; García Lozano 2009). On January 9, 2009, a monument by local sculptor Ricardo Flecha was unveiled to commemorate the catastrophe (see figure 6.8).

Conclusions: Surviving Franco and the Re-networking of Socio-Nature

It does not come as a surprise that toward the end of Franco's life, he was seen as "the great builder of great dams and an example, unique in the world, of a statesman who creates the hydraulic foundations for the progress of his people" (Torán 1971, 314). Frankie the Frog had indeed directed and overseen the complete socio-hydraulic revolution of his beloved fatherland. The revolutionary reordering and remaking of Spain's hydraulic techno-natural configuration welded together discursive, symbolic, geopolitical, socioeconomic processes with new ways of enrolling H_2O within an authoritarian political regime. Achieving this water "activism" depended on the loyal support of a series of powerful interlocked national and international "networks of interests" and coalitions. They often were antagonistic, and required careful massaging and "managing." It is precisely the scalar extension of Spain's political and financial networks within a transforming geopolitics of the postwar world order that facilitated and nurtured the hydro-social transformation of Spain's physical and socioeconomic geography, a project pursued by all possible means until the end of the Franco regime (and beyond). These national and international networks expressed and reshaped power relations in which the voices of the elites

Figure 6.8
Memorial to all those who disappeared in the catastrophe of Ribadelago by artist Ricardo Flecha.
Source: http://commons.wikimedia.org/wiki/File:007370_-_Ribadelago_Nuevo_(8715 658788).jpg (accessed August 21, 2013).

trample over, marginalize, or repress those who dissent. This is particularly acute in a political context in which opposition, even of the mildest kind, comes with serious bodily consequences (imprisonment, exile, forced labor, torture, or execution).

These scalar configurations would both implode and explode after Franco's death and the advent of formal democracy. While the final quarter of the twentieth century showed a perplexing reshuffling of the territorial and scalar configuration of Spain, the Franco legacy proved resilient to change as vested interests and existing power geographies desperately tried to hold on to their powers. The hydraulic engineers and bureaucracy, and the agrarian and boosterist Southern elites wished to complete and perfect the system initiated by Franco. Nonetheless, the stakes would change. New actors, scales, and issues emerged around the hydro-social nexus as formal democracy took root. The demands of regionalists, the actions of environmentalists, the financial might and regulatory order of the European Union (rather than the United States) became increasingly entangled with newly enrolled actants such as birds, wetlands, and river sediments, forging new and different networked arrangements around which radically new socio-environmental and techno-natural projects crystallized.

7 Marching Forward to the Past: From Hydro-Deadlock to Water and Modernity Reimagined

Riau, Riau, Riau, los cables se han cortado![1]
—Solid@rios con Itoiz, 1996

On October 26, 1999, eight activists scaled the construction site of the 415-foot-high London Eye, the landmark Millennium Wheel, to seek attention for the protests against the Narmada Dam Project in India and the Itoiz Dam in Spain's northern Navarra Province, part of the Basque Country.[2] Similar spectacular actions were undertaken at the Vatican in Rome (see figure 7.1), the European Parliament in Strasbourg, and at the Brandenburg Gate in Berlin (İlhan 2009). During the inaugural session of the World Water Forum in The Hague in 2000, activists disrobed on stage to reveal slogans such as "SOS Itoiz," "Don't Privatize Water," and "Stop the Dams" (Bakker 2002, 784). The activists were members of the "Solid@rios con Itoiz" direct action group that, since 1995, used civil disobedience and direct action tactics to protest against the highly contested construction of the Itoiz Dam. The dam, which had been approved in 1985 by the socialist government and finally completed in 2007, is part of a series of hydraulic interventions aimed at expanding irrigation and producing hydroelectrical energy as well as regulating the Irati and Aragón rivers, tributaries of the Ebro. The latter was the cornerstone of an envisaged large-scale water transfer project. For the then Minister of Public Works, Transport and Environment Josep Borrell, who approved the Itoiz project, the dam was part of a larger scheme to "solve once and for all the water problems of Spain" (Beaumont et al. 1997). Immediately after the approval of the project, the *Coordinadora de Itoiz* (the Itoiz Coordinating Committee) emerged as a community-based organization that aimed at halting the project by all legal means possible. The *Coordinadora* became part of a broad-based mobilization that would be replicated in many parts of Spain to organize growing discontent with large-scale hydro-projects (Bárcena Hinojal 1999). A formidable opposition

Figure 7.1
Protest against the construction of the Itoiz Dam, the Basque Country. (a) Cutting the cables, April 6, 1996; (b) scaling the dome of St. Peter's Basilica, Rome, February 25, 2000.
Sources: (a) Solid@rios con Itoiz. (b) http://p2p.kinoki.org/albums/cine_politico/ antiglobalizacion/normal_european_tour_solidarios.jpg (accessed August 23, 2013).

gradually emerged that began to gnaw at the hitherto solid hegemonic national water paradigm and started to rally around a demand for "a New Water Culture." In 2001, the *Coordinadora* assembled eleven thousand protesters in the streets of Pamplona, Navarra's capital city, under the slogan "Stop Itoiz."

Despite the *Coordinadora's* legal appeals and civic action to halt the project, construction continued. The more radical activist organization, "Solid@rios con Itoiz," decided to step up hydro-activism and engaged in a series of spectacular occupations, marches, and sabotages. In the night of April 6, 1996, during the Holy Week, a group of eight activists, equipped with electrical chainsaws occupied the construction site and cut the six 800-meter-long steel cables by which reinforced concrete was transported to the site (figure 7.1). The action was videotaped and several journalists were invited to bear witness of the event. This act of sabotage arrested the building works for eleven months and the activists were sentenced to hefty jail time (Casada da Rocha and Pérez 1996). It also galvanized debates locally and nationally, and the dam protest scaled up to join increasingly louder protests against dams in Europe and elsewhere in the world (Bárcena Hinojal and Ibarra Güell 2001). Moreover, the emblematic Itoiz protest became part of a wider national movement that challenged the orthodoxy of Spain's twentieth-century "hydro-structuralism," a process that had continued unabatedly after the restoration of democracy in 1978.

Indeed, soon after the return to democratic rule in 1978, the debate over national water planning intensified as a plethora of new voices, demands, and conflicts emerged and staged all manner of social actions and mobilizations to protest against the realization of the state's water visions, particularly as the latter basically continued the hydro-structuralism that had marked the Franco era. Indeed, the national "disequilibrium" of Spain's water economy again became highly disputed terrain since democracy was restored in Spain in 1978 and a diverse assemblage of new demands and claims began to creep into the water debate, setting the scene for increasingly strong antagonisms, divergent opinions, and political-ecological conflict. After the return to democracy, the traditional elites continued— at least in the first decade after fascism—with renewed vigor to attempt the completion of the national vision initiated during the Franco years. From the mid-1980s onward, however, the generally accepted national large-scale hydraulic infrastructures and inter-basin water transfer schemes became increasingly challenged, both nationally as well as internationally.

This chapter focuses on the significant socio-political, environmental, and cultural-geographical transformations after the end of fascism as expressed in and by a changing hydro-social landscape. The chapter explores how the hydro-social cycle became one of the central arenas around which Spain's tumultuous political transformations became articulated. The debate over national water planning that raged between 1975 and 2004 chronicles the intense struggle over and contestation of the dominant hydraulic water paradigm. While conservative, engineering, and southern regionalist forces insisted on the completion of the national water grid, a series of new actors emerged (regionalists, localists, environmentalists, parts of the progressive movement, neoliberals, European actors) who began to argue for a radical transformation of the water landscape, one more in line with regionalist, ecological, and market-driven and/or environmental modernization arguments. Indeed, the unfolding of the water conundrum over the past few decades reflects the contradictory dynamics that choreographed a post-Fordist and increasingly Europeanizing, if not globalizing, Spanish economy and society. The state-centered hydraulic paradigm that had structured hydro-social dynamics during most of the twentieth century began to show all manner of cracks and fissures. Indeed, the rescaling of water governance that accompanied the greater autonomy granted to Spanish regions together with the insertion of Spain in a European political and global economic configuration went hand in glove with increasing pressures to neoliberalize the water economy. At the same time, new social conflicts, combining regionalist demands with ecological concerns, further fragmented a once solidly hegemonic socio-political hydrostructural edifice. Even the language in which the water issue was expressed began to change. The emotionally strong, epic, heroic, hyperbolic, nationally charged, and often woolly metaphors that characterized the hydraulic discourse of the past were gradually replaced by a new metonymic enchainment, one that knotted together more sober, techno-managerial, and "rational" metaphors, and more sensitive to diverse social interests and to the new insights gleaned primarily from ecology. This changing discursive tapestry affects of course the evolving narrative around the hydraulic condition. These discursive and metonymic shifts will inevitably infuse the narrative structure of chapters 7 and 8. What some might perceive, therefore, as a rather abrupt change in style compared with the earlier chapters is, in fact, an expression and representation of the changing discursive registers themselves through which the hydro-social complex and its tensions are articulated. Water would indeed again become both material mediator and symbol of changing socio-ecological struggles over the trajectories of modernization.

Democratization and the Endurance of the Hydro-Structural Regime: Retooling Water's Legal Toolbox

In many ways, the transition to liberal democracy after the death of Franco was an exemplary process marked by an apparently very smooth, peaceful, and highly successful process, facilitated, among others, by a national consensus to "forget" the past and move on. The process of democratization was only disturbed briefly on February 23, 1981, by the botched attempt at overthrowing the regime in an operatic coup d'état led by Lieutenant-Colonel Antonio Tejero. Despite the doldrums of a deep recession, continuous political agitation, often-rebellious demands for regional autonomy, and nationalist violence, particularly on the part of ETA,[3] the liberal-democratic transition of Spain had been remarkably painless. Left-wing political parties were allowed again, a new constitution was approved in 1978, and far-reaching autonomy was granted to the regions, in particular to Catalonia and the Basque Country. The international political community, led by the European Community and the United States, as well as the national and international capitalist class supported the shift to a liberal democracy. Authoritarian rule was not any longer legitimate. Felipe González, leader of the Spanish Socialist Workers Party (Partido Socialista Obrero Español, or PSOE), would become prime minister in 1982 and stay in power until electoral defeat in 1996. Under his leadership, the once radical left in Spain gradually moved toward a "Third Way" politics, one that embraced market reform while blunting the sharpest social edges of unbridled market-led capitalism. In 1986, Spain joined the European Community and accelerated the speed of neoliberal reform. Culturally speaking, urban Spaniards shed the yoke of moral restraint, rigid norms, and nationalist fervor. In Madrid and other big cities, exuberant and radically new cultural expressions and forms of social interaction emerged. Known as *La Movida* (the movement/the scene), a new spirit of freedom, transgression of the taboos imposed by the Franco regime, and experimentation with new life styles produced a culturally liberated generation of people who took their cue from moviemakers like Pedro Almodóvar, a new wave of punk-rock bands, and an internationalizing cosmopolitan outlook. While a new cultural and political elite was in the making, "old" Spain continued nonetheless to dwell in its former habits and certainties.

Indeed, a key characteristic of the transition was the restrained attitude toward the bureaucratic strata within the various parts of the state administration. Few, if any, purges took place. Both the administrative structures and the people occupying key posts largely remained the same.

The hydraulic administration continued with its mission to transform the hydro-social landscape on the basis of large-scale and state-led hydraulic interventions and militated to complete further the techno-natural transformation they had been spearheading during the Franco era. However, the 1980s and 1990s mark also an era of heated and persistent water controversies that would gradually, and amid intensifying dispute and proliferating tension and conflict, move the water agenda into new and largely uncharted terrain. Yet, the hegemonic view of the elites remained one that associated economic development, social cohesion, and ecological modernization with a continued emphasis on producing more usable water for those areas, activities, and actors that apparently faced severe shortages. Nature itself entered the array of "actants" that required or was entitled to the right amount of water as ecological concerns began to striate the hydraulic debate.

The pivot of the state's argument remained centered on reinforcing the view that the Spanish hydraulic configuration consisted of deficit and surplus river basins, amplified by a radical mismatch between the agriculturally high-productivity regions of the South that faced high evapotranspiration and water scarcity on the one hand and the less productive but high precipitation regions of the North on the other. These differences had become more accentuated as intensive irrigation and demand from urban tourism in the southern regions required ever-greater quantities of water. Table 7.1 summarizes the state of water affairs in the 1990s and shows the excess (to availability) demand in the southern regions and in Catalonia. This imbalance remained a recurrent theme that sutured the political debate. In fact, the introduction of a more liberal market economy and the onslaught of neoliberalization further reinforced the view that a more balanced water distribution had to be achieved. As Juan Manuel Ruiz contends, a market economy does of itself ensure neither equilibrium nor equity in the distribution of a scarce and geographically unevenly structured resource (Ruiz 1993). In line with the path-dependent state-centered planning and implementation of hydraulic works that had marked the fascist period, centralized hydro-structural control continued—if not intensified and even more entrenched—as the only and necessary strategy to maximize the economic return of water. Julián Campo, minister of public works in the first socialist government after the transition, exclaimed with great pride: "I am going to build more dams than Franco"[4] (cited in Llamas Madurga 1984, 18). Democratization, so the general feeling among hydro-experts was, would finally permit the completion of Franco's hydro-social vision for Spain and fulfill Costa's regenerationist aspirations.

Table 7.1
Total freshwater resources, available resources, demands and water reliability in the hydrological basins of Spain, 2000

River basin resources per capita (m³/hab.)	Total freshwater resources (km³)	Available resources (km³)[a]	Reservoir capacity (km³)	Regulated water (%)[b]	Demand (% of available resources)	Irrigation demand (% of total demand)	Population (millions)	Population per capita	Total resources
North	44.2	6.8	4.4	15	37	42	6.7	6,542
Douro	13.7	8.1	7.7	60	47	93	2.2	6,071
Tagus	10.9	7.1	11.1	65	57	46	6.1	1,784
Guadiana	5.5	3.0	9.6	54	85	90	1.7	3,298
Guadalquivir	8.6	3.6	8.9	42	104	84	4.9	1,755
South	2.4	0.5	1.3	21	268	79	2.1	1,135
Segura	0.8	0.7	1.2	90	253	89	1.4	590
Júcar	3.4	2.0	3.3	58	149	77	4.2	819
Ebro	18.0	13.0	7.7	72	80	61	2.8	6,509
Inner Catalonia	2.8	1.1	0.8	40	122	27	6.2	451
Balearic Islands		0.7	0.3	45	96	66	0.8	785
Canary Islands	0.4		0.4	102	102	62	1.7	241
Spain	**111.2**	**46.6**	**56.1**	**42**	**76**	**68**	**40.1**	**2,728**

Sources: Ministerio de Medio Ambiente 1998, Vallarino Cánovas del Castillo and Garrote de Marcos 2000, and Iglesias et al. 2010, 64.

[a]Surface and groundwater. Overall groundwater contribution is less than 20 percent of total.

[b]Regulated water: percentage of available resources from total "natural" resources.

Soon after the restoration of democracy, debate started over the need to update the existing Water Law of 1879 and replace it with a legal vehicle that would permit the swift implementation of the planned accelerated reorganization of Spain's water geography. By 1985, a new Water Law had been passed that would enshrine and codify the legal framework for water planning for the decades to come. While the nineteenth-century legal vehicle was meant to stimulate (in vain as we have seen) private sector initiatives in the water sector, the key ingredients of the new Water Law of 1985 put the (now rescaled and multilevel) state squarely in the commanding seat and hydrological planning as its cornerstone and master tool, thereby further entrenching the pivotal position of state administrations and their networks in the power geometries of water. In many ways, the new Water Law enshrined legally and codified institutionally what had de facto been the dominant mode of organizing water policy and management during the twentieth century. The 1985 Water Law, therefore, signified more continuity with past practices rather than a new beginning as the state would continue to be the pivotal agent in planning and managing water and financing water works, despite the emerging neoliberalizing dogma pioneered by Margaret Thatcher in the UK and Ronald Reagan in the United States.

The key ingredients of the 1985 Water Law point indeed to the centrality of the state and the affirmation of water as a public good (Gil Olcina and Rico Amorós 2008). First, all waters, both subterranean and surface, were declared to be public domain. Second, the river basin remained as the central management unit. Third, apparently significant autonomy over water matters were devolved to the scale of the autonomous regions. Fourth, the law organized the legal framework for water planning. Hydraulic politics operated under strict national planning at two scales: the National Hydrological Plan (NHP, or Plan Hidrológico Nacional) on the one hand and River Basin Plans on the other. The latter would form the foundation for both the NHP and for the actual planning, management, and infrastructure development in the basin. Finally, the Water Law also introduced a series of environmental and water quality concerns and regulated the system of payments for water extraction, infrastructure use, and the like (Ruiz 1993). Its key objective was "to equilibrate and harmonize regional and sectorial development through increasing the availability of the resource, protecting its quality and rationalize its use" (Gómez Mendoza and del Moral Ituarte 1995). Throughout the final two decades of the twentieth century, preparations for the National Hydrological Plan would become the hornet's nest that aroused intense social and political passions, and accentuated and sharpened social and spatial conflicts that had lain dormant for many years. Moreover, resurfacing "old" conflicts and demands were overlain by newly emerging themes. The NHP would become a pivotal arena that captured the tumultuous transformations of a Spanish world marching into a new era of modernity.

Toward an Integrated System of National Hydraulic Equilibrium: Upscaling Spanish Waters

With the new spirit of openness and democratic debate unleashed by the end of the dictatorship, political controversies became more public and were heatedly debated in a proliferating cacophony of voices, positions, and mobilizations. In particular the hydraulic establishment seized the moment and relentlessly pushed ahead with its vision to complete a century-old dream, the final completion of Spain's reengineered hydraulic landscape. Throughout the 1980s and early 1990s, the preparations for the NHP progressed steadily in the design and policy rooms of the state hydraulic engineers and administrators. The Water Law of 1985 gave a further impetus and urgency to its completion, spurred on by a socialist government that had fully embraced at the time the old regenerationist ideals and vowed to solve finally Spain's water problems.

The *Revista de Obras Públicas* insisted with characteristic technocratic zeal on the importance of the NHP and the *transcendental* necessity to complete the long desired large-scale water transfers (Martín Mendiluce 1996), together with the further expansion of new dam constructions. In close collaboration with the Dirección General de Obras Hidráulicas, the highly corporatist state body that had traditionally been mandated with the development of the state's water resources and maintained excellent and close relationships with the broader hydraulic network of interests built up after decades of unchallenged hegemony over the planning and implementation of hydraulic works, the engineers, this "brotherhood of concrete" (Martínez Gil 1999, 109), had produced a "stagnation of hydrological thought" (Díaz-Marta 1996) as they stubbornly insisted on a singular set of objectives and means to achieve them. Immediately after the end of the dictatorship, the engineering community indeed staked out its claims in the *Revista de Obras Públicas* for a further integrated planning of water resources, under national control and with a strong rational-technical emphasis, and insisted on the need to build more dams and to regulate the national hydroscape in an integrated manner (see, for example, Palancar Penella 1976; Couchoud Sebastia 1977; Sancho de Ybarra 1977).

The old idea of major water transfers was regurgitated with a vengeance (Lafuente 1990). The increasing number of contributions that eyed explicitly to intervene in the water policy domain is all the more telling as the Corps of Engineers had systematically refrained from making explicit policy—let alone political interventions—during the dictatorship. For example, Mariano Palancar Penello, then director of the Guadalquivir River Basin Authority, insisted on the need to build more mega-dams as "dams provide water" and since "water supports life," dams are a good thing, although he added in a leading manner that "environmental impact studies need to confirm" the positive impact of dams (Palancar Penella 1979, 1021). This emphasis on building even more dams was fully endorsed by the scientific engineering community (Alvarez and Del Campo 1979), whereby the hydrological cycle "changes from being the mere expression of a physical reality to become the unifying element of hydraulic politics" (González Paz 1994, 33). In 1982, Martín Mendiluce and Torres Padilla summarized the hydraulic task ahead as follows:

The attention to quantitative problems of water . . . must be permanent. It is not possible to rest in the task of increasing the availability of hydraulic resources. The Spanish experience is well known; after a period of intense activity (1950–1970), a short break seems well deserved (1970–1980); but this attitude has brought as a consequence shortages that require today to strengthen the activity to levels superior to those achieved during the periods of greatest apogee.[5] (Martín Mendiluce and Torres Padilla 1982, 221)

In 1996, in the midst of an acrimonious electoral campaign that at least partly focused on national water planning, the need to accelerate the building program of new dams was reiterated. All this is a sure sign that Franco's wet dream for Spain was still alive and its main protagonists working away like beavers at its final realization (Martín Mendiluce 1996). Nonetheless, the obvious feeling in the hydraulic community that the old slogans and demands had to be repeated and reinforced also suggested that the writing was on the wall, that dark clouds were gathering over the once solidly hegemonic hydro-social vision.

The Draft Proposal for the National Hydrological Plan was finally published in April 1993 and hit the water community like a bombshell (Ministerio de Obras Públicas y Transportes 1993a). Although a plan of "concrete and bulldozers" was largely anticipated, the actual planned proposals went way beyond even the wildest imaginations (Llamas Madurga 1993, 99). The vision explored in the draft NHP was truly gargantuan in scale and scope. The planned works would transform radically and fundamentally the national hydro-social configuration. The draft NHP took up the notion of and envisaged the completion of an Integrated System of National Hydraulic Equilibrium (Sistema Integrado de Equilibrio Hidráulico Nacional, or SIEHNA). Already prefigured during the Franco era, SIEHNA would finally physically upscale what had been demanded for decades, namely expanding the management of Spain's mainland waters from primarily a river basin perspective to the scale of the entire national space. The doctrinal core of the draft NHP was, like the one before, the thesis of the natural hydraulic disequilibrium, with a dry Spain on the one hand and a wet Spain on the other. Paradoxically, according to the draft NHP, Spain has more water per capita than Denmark. Moreover, the torrential nature of the hydrological system was a further defining sign of a country in which "the Creator had made a mistake." The plan argued that the identified deficit of 3,000 hm^3 in 1993 would inflate to 6,000 hm^3 per year in 2002 and to an incredible 9,000 hm^3 in 2012. "An apocalypse would happen if this was not corrected in time," something that was further fueled by daily disaster stories in the media because of an epochal drought that ravaged irrigation agriculture and curtailed urban water supply. Hundreds of new dams and fourteen major water transfers would rectify the socio-hydraulic imbalance. The SIEHNA system would become "a great centralized and absolute power apparatus to manage the distribution of excess waters" (Llamas Madurga 1993, 110–111).

The draft NHP proposed to interconnect all river basins with each other with an eye toward transferring significant volumes of water from the Ebro, Duero, and Tago rivers (with their presumed surplus water) to the "deficit" basins of the South, the Levant, and Catalonia. In a first phase, with an anticipated completion by 2002, 2,450 hm^3 of water would be transferred. By 2012, the national project would be completed with a total volume of 3,768 hm^3 per year of transferred water. Figure 7.2 summarizes the reimagining and planned reengineering of Spain's water balance for 2002 and 2012. If implemented, this project would easily be the largest infrastructure project ever undertaken in Europe. The total estimated investment cost for carrying out the draft NHP was estimated to be 3,600 billion pesetas in 1993, equivalent to 21.6 billion euros (at 2001 exchange rates). This would be about three times more costly than the Channel Tunnel, the largest infrastructure project in Europe to date. At the same time, a "State Organization for National Hydraulic Equilibrium," a kind of super-basin management organization, would be established to organize and police the national water grid (Gil Olcina 1995).

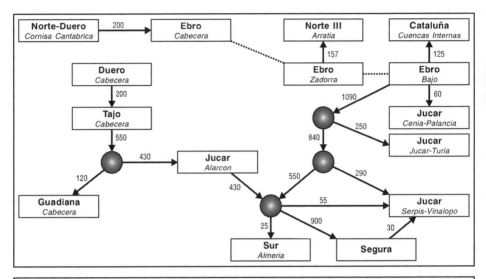
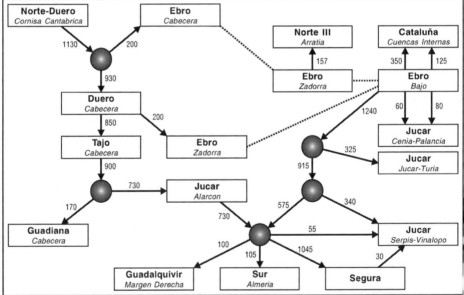

Figure 7.2
Integrated system of national hydraulic equilibrium: Planned inter-basin water transfers, 2002 and 2012.
Source: Ministerio de Obras Públicas y Transportes 1993a, Grafico V.2.

For Borrell, who was politically responsible for the draft plan, "rationality" and "solidarity" were the main guiding principles underlying the NHP and this would establish "an authentic politics of territorial structuring (*vertebración*[6]), social cohesion and environmental conservation" (Borrell Fontelles 1993, 5). The traditional modernizing elites, spearheaded by the Corps of Engineers, believed strongly that the regenerationist vision "to crisscross the country with an arterial hydraulic system to mitigate the heat and satisfy its thirst" (cited in Gil Olcina 2002, 18), pioneered by Joaquin Costa at the turn of the twentieth century, would finally be realized and solve Spain's water fever forever. The *Revista de Obras Públicas* published a special issue to discuss and celebrate the completion of the draft plan. In his contribution, the director general of the DGOH insisted, in classic regenerationist vein that "water should be one of our fundamental concerns, and hydraulic politics should take a preeminent place in territorial politics"[7] (Baltanás García 1993, 7). In one stroke, all remaining hydraulic problems would be swept away: the South would receive enough water to secure a safe path to further water-intensive agricultural and economic modernization, illegal overexploitation of coastal aquifers would stop as the arrival of the North's excess waters would annul the need to tap into unsustainable aquifer extraction. The accompanying presentation of the draft NHP's motives extolled the virtues of the plan and defended its mission with a turn to reviving the classic early twentieth-century regenerationist rhetoric:

In addition to all the economic, social and legal changes, the same hydraulic problems that at the time founded and still constitute today the basis for the pursuit of a resolute intervention of the state in the public water domain in the form of hydrological planning. . . . This law [the draft NHP] then contains a projection of the transfers of water resources between river basins oriented to and guided by an objective of great ambition: it is about laying the foundation that will permit, for once and forever, solving the manifestly uneven distribution of Spain's water resources.[8] (Ministerio de Obras Públicas y Transportes 1993b, 3, 11)

The preamble concludes poetically that the "final solution" presented in the NHP "on the eve of the beginning of a new century" completes "the realization of Costa's old dream for Spain" whereby "the Ésera [a river in Costa's native Aragón[9]] and many other Éseras will run through the skin of Spain and its clear waters will be, recalling the poetic words of Costa, its blood, its dew and its gold, the path to liberation and collective wealth. For this, the National Hydrological Plan is prepared"[10] (Ministerio de Obras Públicas y Transportes. 1993b, 20). Indeed, one hundred years of mobilizing the formidable powers of water in the name of progress, development, and Spain's national collective well-being would at last find its ultimate resolution, a collective vision that cements the solidarity of all of Spain in the national redistribution of its tamed waters. The main driver remained of course the agricultural sector and its insatiable demands for more water to irrigate the dry lands of the South, but now complemented with a booming Mediterranean tourist industry and its escalating thirst for more water. The plan envisaged an expansion of the irrigated area with more than 600,000 hectares, bringing the total to almost 4 million hectares. Of course, the draft NHP also pays obligatory tribute to newly emerging issues such as the need to safe water, improve its efficiency, and recycle residual waters.

However, the master planners had not reckoned with the depth and breadth of the changing mood that had engulfed Spain after the dictatorship. Very soon after the publication of the draft NHP major controversies began to arise. Even from within the hydraulic engineering community, the first voices of dissent were raised against the single-minded emphasis on supply-side measures through the construction of large hydro-infrastructures. In particular, issues of groundwater management, environmental and ecological concerns, re-use of water, and demand management began to make inroads into the debate on the preparation of the plan from the part of a new engineering generation that had begun to formulate hesitantly what later would be become known as ecological modernization (Díaz-Marta 1993; Sahuquillo 1993). Manuel Llamas considered the creation of an Integrated System of National Hydraulic Equilibrium to be motivated by a desire "to reinforce even more the Orwellian figure of 'Big Hydraulic Brother' who will decide the destiny of all of Spanish waters" (Llamas Madurga 1993, 101). Moreover, having the lever on water transfers permitted the national government to use water as quid pro quo in negotiating agreements with regionalist parties that had become increasingly important to form national coalition governments. Nonetheless and despite the emergent rumbles of protest, all seemed now set to march confidently into the future as early twentieth-century imaginaries would finally materialize at the dawn of the twenty-first.

Deadlock!

However, for the next twenty years, the NHP would become one of the central political controversies in Spanish political life, resulting in a deadlock that would last well into the early twenty-first century. After its completion, the draft NHP was sent to the National Water Council (NWC) for consideration. The NWC is the highest consultative body for water affairs with representatives from the national, regional, and local governments, the river basin authorities, irrigators, and nonprofit interest groups (unions, employers, environmentalists). The NWC received 1,143 objections to the draft plan from a wide range of social, political, and economic actors. In July 1994, the NWC approved the draft with some recommendations for change, after which the Department of Public Works revised slightly the draft NHP in 1994. All representatives of the receiving water regions voted in favor of the plan, all those from the "donating" regions voted against, as did the environmental organizations and the communities of irrigators (Gil Olcina 2002, 34). The total volume of planned final water transfers was reduced slightly to 3,210 hm^3 per year (down from 3,768 hm^3), but remained the cornerstone of the plan. The total planned investment cost was increased by more than 60 percent (Gil Olcina 1995). The draft NHP was now ready to be sent to the *Cortes* (parliament) for approval. However, the process of approving the draft plan unraveled quickly. After the elections of 1992, the Socialist Party had to form a coalition government with the support of regionalist parties, but its once-hegemonic political position was crumbling quickly. *El Congreso* (the Lower House) accepted unanimously a motion put forward by the oppositional conservative Partido Popular (PP) that stipulated that the government had to present, together with the final version of the NHP, a National Irrigation Plan that took into account the European reforms of the Common Agricultural Policy and the stipulations of the General Agreement on Tariffs and Trade (GATT)[11] as well as a "National Plan for Saving and Re-using Water." The Senate in its turn decided in September 1994 that River Basin Plans had be to completed and approved before the NHP could be presented to the Senate (Gómez Mendoza and del Moral Ituarte 1995). Only the socialist fraction, in an attempt to defend the NHP, opposed the motion. None of these necessary requirements were fulfilled at the time. Both initiatives clearly aimed at slowing down, delaying, if not blocking, the approval of the NHP (Cimadevilla Costa and Herreras Espino 1995). When the Partido Popular came to power in 1996 after thirteen years of uninterrupted socialist-dominated governments, the NHP was in tatters and the new government had to pick up the pieces (Pérez Díaz and Mezo 1999).

In addition to the political parties, civil society groups raised their voices. The umbrella organization of the environmental groups (CODA) was squarely against the water transfers (Coordinadora de Organizaciones de Defensa Ambiental 1994) and the World Wildlife Fund was highly skeptical because of the anticipated negative environmental effects of implementing the plan (World Wildlife Fund 1994). The unions of small farmers were resolutely against the plan as they believed water transfers would intensify territorial inequalities and favor large landholders in the South (Coordinadora de Organizaciones de Agricultores y Ganaderos 1993, 1994; Union de Pequeños Agricultores 1993); large construction companies were unsurprisingly great advocates of the plans as were the large southern irrigators. Both were set to benefit greatly from the planned new infrastructure that would continue to socialize the investment costs while channeling the financial benefits to a very select group of socioeconomic actors. The leading economic sectors insisted on the need to introduce greater market efficiency and demanded more integrated and coordinated planning (Consejo Económico y Social 1996). Critical academics raised all manner of social, economic, and environmental objections and debated at length various possible alternatives such as desalination, more efficient water use, or altering the traditional water-hungry development model (Arrojo Agudo 2001a,b). Whereas the Franquist regime could easily silence dissenting voices, the democratic transformation had unleashed an unparalleled explosion of arguments advanced by a wide range of interests, advocating often-irreconcilable positions. This cacophony of voices propelled water to an issue of great public concern, marred by all manner of controversies and virtually nonnegotiable positions, produced an extraordinary and often emotional debate, but also a stalemate that froze the hydrological terrain in a political deadlock, what Martínez Gil called a "ceremony of confusion and non-governance" (Martínez Gil 1999, 115). It also signaled the demise of the "traditional hydraulic policy community" and of the monopoly position of the DGOH in matters of water resources development (Pérez Díaz, Mezo, and Álvarez-Miranda 1996). Ironically, turning water into a matter of public concern hit the limits of traditional political and managerial structures and practices.

The pretensions of the NHP to structure the entire national hydrological space in a redistributive manner, a vision underpinned by the nineteenth and twentieth century construction of a united and indivisible Spanish nation and materialized through technical and political infrastructures of all kinds, had begun to crumble as political configurations were redrawn and new social demands and aspirations asserted themselves. In the process, water became reimagined, inaugurating new forms of political performativity. The common sense view that had permeated a century of hydraulic interventions was punctured and new arguments and visions vied for the formation of a new common sense, away from old-style "hydraulic structuralism."

From Deadlock to Gordian Knot: Neoliberal "Glocalizing" Waters

After the inauguration of the democratic transition, the ruthlessly quelled quest for far-reaching autonomy, primarily by the Catalans and the Basques, during the Franco era, and the rise of revived regionalist aspirations found its constitutional resolution in the establishment of Autonomous Communities (*Comunidades Autónomas*) in 1978. Today, Spain has a quasi-federal structure with directly elected regional parliaments (seventeen in total) and legislative powers for a wide-ranging set of issues (including the management of regional waters). The regions manage over 40 percent of the Spain's public expenditure. While originally the impetus came primarily from the fast-track regionalization demanded by Catalonia, Galicia, Andalusia, and the Basque Country, this was quickly followed by more assertive regionalist identities and political demands for autonomy from the other regions as well, particularly, but not exclusively, in the domain of water. A host of new water institutions at a range of interlocking geographical scales gradually replaced the earlier top-down state-centered hydraulic politics. The Water Law of 1985 stipulated that river basin authorities (RBAs) whose geographical base falls wholly within an autonomous region are under the authority of the Autonomous Communities (which created regional water agencies), while RBAs that straddle one or more region remained under national control (Vera Aparicio 2009). Of course, most of the key river basins implicated in the water transfers were transregional and, therefore, under national control. Nonetheless, the greater regional identification cut through traditional nationally organized party divides and water would quickly become one of the pivotal axes around which regional political arguments would become articulated. Inter-regional conflict increasingly overlaid traditional left-right conflicts and the proposed large-scale water transfers became a veritable Gordian knot in the management of territorial political tensions, particularly between the water-ceding and the water-receiving regions.

In the proposed NHP, the Autonomous Community of Aragón was clearly the most affected by the planned Ebro river diversion. Already in June 1992, in the run-up to the completion of the draft plan, the Aragón regional parliament fired the opening salvo for years of intensified hydro-social conflict as a regional cross-party Water Pact (*Pacto del Agua*) was agreed upon. Burying long-standing party squabbles, a territorial alliance was forged that insisted that the rights and interests of the ceding river basins should take priority over those of the receiving basins. The pact claimed that at least 11,200 hm^3 of Ebro water was required for "local" use and that the financing earmarked for hydraulic works should foment development in Aragón. The pact took a clear anti-transfer position (Gil Olcina 2002, 32–33). The region of Castilla and Léon, which was affected by the proposed transfer from the Duero river, agreed to it "provided the transfer would be placed close to the Portuguese border" (34), knowing full well that this would escalate already tense hydro-political relations with Portugal. The Duero (together with the Tajo) provides most of the surface water for Portugal as they both drain into the Atlantic Ocean.

The envisaged national water solidarity had begun its long eroding march, despite the insistence of the Secretary of State for Water Politics and the Environment in 1993 that "if the need to connect basins nevertheless arises inexorably, it has to be taken into account that river basins are no more than an administrative division, and that water is of all Spaniards, and this is reflected in the Water Law"[12] (Gil Olcina 1995, 24). During the subsequent decade, the interregional water wars would further intensify and tighten the Gordian knot of Spanish hydraulic politics as water had now been decisively mobilized in territorial rivalries, regionalist politics, budding environmental sensitivities, and their accompanying electoral tactics. Moreover, the severe drought that had affected much of Spain during the first half of the 1990s and which led to serious water stress, both for agricultural and urban uses, fused seamlessly with the intensifying conflict over the direction of the nation's water politics and further propelled the material performativities of H_2O and its socio-physical dynamics to the top of the political and social agenda.

While much of the water debate focused on regional and national concerns and interests, a silent revolution was unfolding in Spain, one that would engulf water too. Ever since the deep economic crisis of the 1970s that was usually referred to as the crisis of the Fordist development model and especially after the return to democracy, the hegemonic forces in Europe and beyond drove a class politics that increasingly shaped the political-economic configuration in the image of the capitalist elites and their global neoliberal market project. This shift was emblematically enshrined in and facilitated by the transformation of the Spanish (and other) traditional socialist left. While both communist and socialist left-wing parties had been historically deeply committed to a statist project, the fall of Franquism in Spain and the crumbling of eastern European socialism made state control and intervention more suspect, while the social-democratic forms of Fordist state intervention were faced with fundamental political-economic problems. In addition, the project of European integration was shifting from a political project pursued by economic means (the establishment of a common market) to an economic project of neoliberalization and integration into the global economy pursued by political means (the successive European treaties). Soon after the return of democracy, the Spanish socialist left had indeed embraced a "Third-Way" political perspective, as one of the first social-democratic parties in Europe and long before Tony Blair would adopt it in the UK and popularize its principles more globally. Decidedly rejecting the traditional left dictum of state control and anti-capitalist reform while embracing market reform with a social consciousness as its guiding principles, the edifice of left-wing politics in Spain began to shift decidedly in a way that would unashamedly embrace neoliberal globalization. Spearheaded by a new generation of emblematic and inspiring leaders, who had not grown up in the resistance against Franco, they began to assert themselves and dominate the political landscape. Pasqual Maragall, for example, became the charismatic socialist mayor of Barcelona who, from 1982 onward, would manufacture the Barcelona miracle, the transformation of a sedate Mediterranean city into a thriving global tourist attraction, spurred on by an endless series of iconic global cultural-economic projects. Pasqual Maragall had, of course, learned his trade in the United States where he had been looking at Mayor William Donald Schaefer's iconic Baltimore waterfront revival while a research fellow at the Centre for Metropolitan Planning and Research at my alma mater, the Johns Hopkins University. Equipped with these new neoliberal recipes of entrepreneurial urban revival, he returned to begin the emblematic transformation of Barcelona. At the national level, the Andalucian Felipe González had single-mindedly maneuvered the PSOE away from its historical Marxist roots. During the party's twenty-seventh congress in 1979, González resigned as party leader, because the party did not follow his and his allies' demands to abandon its "Marxist" character. A few months later, an extraordinary twenty-eighth congress was called. In a political turnaround, the PSOE then agreed to move away from Marxism and González was reelected as the uncontested leader. He would remain in post until the party's electoral defeat in 1996, after an unprecedented four successive electoral victories. He resolutely embraced a politics of globalization by pursuing accession to the European Economic Community in 1986, joining NATO in the same year after a referendum in which he defended Spain's membership to the organization, and supporting the 1991 Iraq War.

González became Spain's globalizing wonder boy. During the long mandate of the Socialist Party, the state's involvement in social and economic matters was gradually reduced, marked by a slimming-down and rationalization of state bureaucracies, the sale or part-privatization of state-owned companies (also vigorously pursued by the subsequent conservative government), privatizing industries like telecommunications, SEAT, Repsol, Endesa, and Argentaria (Gamir 1999), industrial restructuring, and the reduction of state subsidies to loss-making operations (Holman 1996; Murphy 1999). At the same time, however, the socialist government extended and generalized welfare provision in an attempt to massage the excesses of market reform and to maintain the support of its popular base. Managing these tensions resulted in severe internal conflicts. Indeed, much of the impetus derived from the need to reform state spending and bring down the deficit while maintaining social cohesion and political support. The contradictory demands of these Third Way politics resulted in a situation in which both debt and budget deficits increased as a percentage of GDP, sowing the seeds of the unprecedented economic and financial crisis that would hit Spain in late 2008.

The gradual erosion of the PSOE's electoral position resulted in a regime switch in 1996, when the conservative Partido Popular, under the leadership of José Aznar, won the elections and would stay in power until 2004. While the reign of the PSOE was still marked by a state-centered process of hesitant, incoherent, and fragmented neoliberalization, Aznar would fully embrace a radical anti-statist, privatizing, and globalizing strategy, thereby wholeheartedly supported by the American neo-conservatives and the Transatlantic neoliberals (Chari and Heywood 2009). The slimming down of the state, the tight fiscal regime, the financialization of the economy, and the imperative of neoliberal European integration would be etched into the restructuring of the hydro-social cycle. Indeed, hydro-politics would express the tumultuous and conflicting social struggles that marked the process of neoliberalization and account in part for the deadlock that characterized much of the 1990s hydraulic policy landscape.

While the 1985 Water Law, passed by the socialist government, still very much expressed the old regenerationist dreams and aspirations, the neoliberalizing process with its contradictory dynamics of increasing interregional conflict combined with Europeanization and market-led globalization, rendered its implementation increasingly improbable. While much of the Orwellian designs underpinning the draft NHP of 1993 were supported by traditional state-centered socialists, the subsequent more mercantilist neoliberal course of the government turned this "structuralist hydraulic project" into a white elephant that ran increasingly against the newly constructed dogmas of the virtues of private enterprise, flexible specialization, and market-led development. First, economists and assorted activists began to question the long-held assumption that the contribution to national wealth largely exceeded the cost of state-financed hydraulic works.

Introducing a mercantile market logic into water resources management was believed by many, including some environmentalists, to weed out suboptimal use of water, maximize economic efficiency, foster water saving, and benefit environmental concerns (del Moral Ituarte et al. 2003). Second, the Treasury was highly critical of the exuberant state spending envisaged by the draft NHP, particularly given the significant cuts imposed by budget downsizing and demands for marketization/privatization in the run up to achieving the European Union's convergence norms after the 1992 Maastricht Treaty—a necessary precondition to be allowed to join the new common currency planned for 2002. A hydraulic system that recovered only about 0.2 percent of the replacement cost of public investments in dams and canals was no longer deemed sustainable (Ayala-Carcedo 1999; Ministerio de Medio Ambiente 1998, 555). Finally, in view of the expansion of the European Union eastward after the fall of the Berlin Wall, the anticipated reduction and eventual end of the European Union's Structural and Cohesion Funds from which Spain had generously benefitted to finance infrastructure and regional development projects of all kinds, and on whose support the implementation of the NHP would depend, also jeopardized the feasibility of public hydraulic projects. Only private financing of water works offered a possible way out (Bakker 2002, 776–777). The *mercantilicazión* (to use Karen Bakker's term) of Spain's water became indeed an emblematic, albeit contradictory, process marking a significant shift in Spain's hydraulic edifice. However, the process of attracting private investment that had remained elusive in the late nineteenth century was certainly not easier in the late twentieth century. The lack of profitability, the high upfront cost, the long amortization periods, and the uncertain long-term returns of large-scale infrastructure rendered private investment a difficult proposition (Turró 1996). Ingenious recipes had to be found that attracted private investment with state support, a process initiated by the PSOE but further entrenched during the later conservative regime.

In 1996, a reform was passed that permitted the government to create *sociedades estatales* (state companies) that operate under private law (Vera Aparicio 2009). The newly created state water companies were charged with facilitating and attracting private finance for the construction and management of hydraulic works selected by the state (del Moral Ituarte et al. 2003) and promoted as major instruments for the financing of new projects in order to "break with the historical tendency in which the state was solely and exclusively responsible for Hydraulic Works" (Embid 1998, 288). As the Secretary of State for Water and Coasts Benigno Blanco Rodríguez affirmed: "This instrument permits opening new modes of private sector participation in the development of hydraulic infrastructure. This permits a greater responsibility of the private sector in the financing of infrastructures of water supply and the exploitation of water resources"[13] (Blanco Rodríguez and Rodríguez González 1999, 7).

The more optimistic voices believed that such participation would accelerate investment while permitting state funding to be diverted to less profitable activities (like irrigation, for example) (Valero de Palma Manglano 1999). Of course, the establishment of state companies also permitted all manner of other hitherto public water companies (like municipal water supply) and other sectors to operate like private companies and draw on market relations and forces more than on standard bureaucratic procedures and exclusive state funding.

A much greater step forward toward full marketization and commodification of water was enshrined in the amendments to the Water Law in 1999. These amendments would change the concessional water arrangements and establish an institutional and legal framework for a further full marketization of water. Three main components pointed in that direction. First, it regulated the marketization of desalinated water. The conservative government had initiated the construction of a limited number of desalination plants (see chapter 8 for a more thorough analysis of this move to the sea). The new law stipulated that desalinated water, from the moment of desalination to its point of return into "natural" and "continental" water flows, was owned and managed as a private commodity. In other words, the collective part of the hydro-social cycle became socially and economically split into respectively a private and a public component. While the sea is a common resource, its water after extraction would be private until it returned to the "natural" flow, when it would become a public resource again. The objective is of course to incentivize private sector interest in desalination (Embid 2002; Garrido 2006).

Second, the amendment permitted the establishment of water markets that, under certain conditions, allowed concessionaries to sell water to those who required it. Although water *rights* cannot be transferred—they are still allocated by the state—water right holders could transfer the *use* of the water to others. The establishment of water markets in the law was undoubtedly inspired by earlier neoliberalizing water experiments in Chile and California (Arrojo Agudo and Naredo 1997; Bauer 1998; Haddad 2000; Budds 2004). Third, the amended law also sanctioned the establishment of water banks. The latter could be established by the state in exceptional circumstances (such as droughts) (Giansante, Babiano, and del Moral Ituarte 2000; Saurí and del Moral 2001). All three legal inscriptions contributed in important ways to facilitating the commodification of the hydro-social cycle and to render water allocation more flexible and dynamic, and make it conform to market rule. As Karen Bakker notes:

> The reforms of the Water Law are ultimately intended to reduce if not eliminate the state subsidy to the hydraulic network, to allocate water to (economically) highest value uses, and to incorporate full costs to society and the environment in water charges. Rather than impose these changes directly via "command-and-control" type measures..., the administration has elected to implement market-based mechanisms. (Bakker 2002, 780)

Of course, all this was in line with the European Union's demands and the preparations of the European Union's Water Framework Directive that demanded full-cost recovery of water projects (as well as increased participation of all relevant stakeholders) (Kaika 2003; Kaika and Page 2003; Page and Kaika 2003).

Technological argument and the principles of "good management" practices were increasingly mobilized to massage and deflect the intensifying tensions and contradictions that marked the hydraulic landscape. These techno-managerial conduits permitted negotiating a fine balance between nurturing the demand for the production of more usable water on the one hand and fulfilling the imperatives of the market, accelerating the commodification of water, and supporting market-led water infrastructure development on the other. While the socialist government tried to continue the twentieth-century hydraulic paradigm in parallel with hesitantly introduced market-based reforms, the conservative government intensified the already acute tensions between those that desired a continuation of the old system and those who considered full-blown marketization as a wedge to transform the hydro-social configuration in an environmentally modernizing way. The latter aimed at enhancing environmental standards through the introduction of full economic costing and market-conforming mobilization of water. Needless to say, these contradictions would further intensify as the process unfolded and ultimately came to a head as Spain marched into the new century.

The Impossible Compromise: A New Plan, a New Century!

The neoliberalizing drive of the conservative government had not alleviated the discussions and disputes over the National Hydrological Plan that the new government had inherited from its predecessors and which the 1985 Water Law had made mandatory. In 1998, the National Irrigation Plan and the river basin plans had finally been approved, eliminating the obstacles that parliament had put in the way of the NHP's approval in 1994 (Ministerio de Agricultura, Pesca y Alimentación 2008). In addition, the new water administration, now part of the Ministry of the Environment, published in 1998 the *Libro Blanco del Agua en España* (The White Book on Spanish Water) (Ministerio de Medio Ambiente 1998), a hefty volume that presented the state of Spain's complex hydro-social configuration. Much to the dismay of a growing number of dissenters to the traditional "structural" hydraulic politics, the White Book still concluded its presentation with a discussion of the disequilibrium between "deficit" and "surplus" regions. The White Book ignored completely the increasingly vocal affirmations that neither "surplus" nor "deficit" is structured solely by the "natural" distribution of rainfall and the orographic structure of the country, but rather by the complex historical-geographical production of a hybrid hydro-social landscape that had turned the Southern regions into endlessly water-gulping irrigated lands. It remained taboo to acknowledge how an intensifying hydro-agrarian development model, combined with booming speculative real-estate development, produced a structural socio-naturally produced scarcity of water and rallied most social and political forces around relentless demands for more water (Gómez Mendoza and del Moral Ituarte 1995; del Moral Ituarte 1996). Despite urgent calls to reconsider the territorial political-economic dynamics of this uneven development, the White Book repeated the old discursive frame (Martínez Gil 1999; del Moral Ituarte 2009).

In a frantic sequence of plans, new laws, and changing policies, José Aznar's government finally presented a revised National Hydrological Plan in 2000. The *Cortes* approved this final installment of a long odyssey in July 2001 (Ministerio de Medio Ambiente 2000a,b). This final plan really tried to square the circle, thereby satisfying no one and accentuating even more clearly the escalating contradictions that sutured the water conundrum (Arrojo Agudo 2001a). While still reveling in a constructionist logic that anticipated dotting 200 more dams over the Spanish landscape and advocating a logic of continuing transfers, the ambitions of the plan had been scaled down significantly compared with the provisional 1993 project, while water pricing and environmental concerns took a much more prominent place. The most important change was undoubtedly the reduction in anticipated water transfers. The Duero transfer was scrapped completely and the total of planned transfers to the coastal regions was reduced to a still formidable 1050 hm^3 per year, which would be delivered by means of a new infrastructure project that would take "excess" waters from the Ebro River to the South and to Catalonia. The Júcar and Segura river basins in particular would receive significant volumes and Almería and Catalonia would also be among the recipients of the Ebro's "surplus" water (see figure 7.3). The transfer to Catalonia would be 170 kilometers long, the one to the south would cover a total distance of about 750 kilometers; 330 km would be in an open canal, 100 km in a tunnel, 400 km in pipes, and 90 km in aqueducts (Menéndez Prieto n.d.). Combined with the newly planned regulating dams, this transfer was still the plan's backbone. However, the price of water to be paid by the receiving areas would be higher than for the existing Tajo-Segura transfer and Aragón, the most affected region, would receive "generous" economic compensation for the "loss" of part of "its" water. Still, arguments of "national solidarity" and "territorial equilibrium" were the discursive vehicles to defend the necessity of this Ebro transfer. However, the plan that had been coveted for two decades became immediately mired in dispute and controversy. The plan's future would indeed prove to be highly uncertain.

Turmoil at the Turn of the Century

Mass protests immediately followed the approval of the NHP. The Platform for the Defense of the Ebro (*Plataforma en Defensa del Ebro*) was one of the most vocal and effective organizations that militated actively against the planned Ebro transfer and for a "New Water Culture." It ultimately managed to mobilize large and heterogeneous segments of the Spanish population (Pons Múria 2003). This movement brought together an often-uneasy alliance of environmentalists, regionalists, socialists, and local activists. In their heterogeneous claims and demands, the protesters mobilized a diverse set of human/nonhuman issues: the rights of fish and fishermen, the fate of river sediments and sea shorelines, the life of birds and plants, the preservation of wetlands, the protection of local livelihoods and regional cultures, the concern for an ecologically sound environment, and the need to preserve water in the name of sustainability, water's mythical values, and nature's or people's rights to water (Arrojo Agudo 2001b, 2004). The activists' primary target was the Pharaonic plan, the centerpiece of the new National Hydrological Plan, to transfer large quantities of "surplus" water to the "deficit" basins of the semi-arid Southeastern regions of the Levant on the one hand, and to Barcelona on the other.

For example, on March 10, 2002, an estimated half a million Spaniards marched through the streets of Barcelona protesting against the National Hydrological Plan that had been recently approved by the conservative government of José María Aznar and his Partido Popular. Earlier, in 2000,

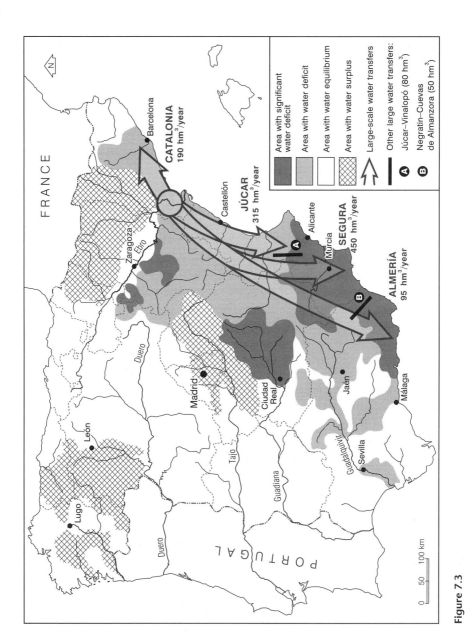

Figure 7.3
National Hydrological Plan: "Surplus" and "deficit" water areas and the planned Ebro transfer, 2000.
Sources: Based on Ministerio de Medio Ambiente 2000a,b and Jefatura del Estado 2001.

more than 400,000 people had gathered in Zaragoza and another 250,000 in Madrid and Barcelona to voice their discontent. These demonstrations were among the largest ever seen in Spain, and protests spread to many other towns and cities around the country (see figure 7.4). Of course, irrigation-based farmers, urban developers, golfing enthusiasts, and political elites of regions that would receive the "new" water raised their voices in dissent against this protest, and manifestations in support of water transfer schemes were organized in cities like Almería and Murcia. The agricultural elites, the urban boosters, and the tourist bosses claimed they desperately needed the extra water to sustain the Levantine regions' development. This "water" struggle was also taken to Brussels, when the "Blue March," with ten thousand activists, arrived at the headquarters of the European Union in September 2001 after a 1,500-kilometer trek from Spain to Belgium, demanding that the European Union live up to its own water rules—as enshrined in the Water Framework Directive (see Kaika 2004)—and stop the Spanish NHP (NN. 2001).

Even Pope John Paul II joined the controversy. In a speech to the Spanish Bishops on January 24, 2005, a few weeks before his death and shortly after the contested water transfer was suspended by the socialist government,

Figure 7.4
March against the National Hydrological Plan, Valencia, 2002.
Source: http://elpais.com/elpais/2002/11/24/actualidad/1038129420_850215.html (accessed August 21, 2013). Copyright Agencia EFE, Madrid.

the Pope insisted that it is a Catholic principle to distribute water justly: "Some areas live with abundance while others have severe shortages. . . . In some parts one lives through social confrontation for a natural resource: water is a common good that should not be squandered or the duty of solidarity to share its use be forgotten. Wealth cannot be monopolized by those who possess it"[14] (John Paul II 2005). The Partido Popular and southern (Catholic) elites of course welcomed this unashamedly political statement that was clearly and directly targeting the demands of the ecologists and others protesting against the water transfer.

So, the twentieth century ended very much where it had begun. Multiple actors and interests were pulling in a variety of, and often directly opposing, directions; major transformation and changes had happened and were clearly still under way. Modernization, while still on the agenda, was making a new turn into largely uncharted terrain. It would have to be more sensitive to ecological concerns, navigate a variety of old and new but often conflicting positions, and take account of local demands while tuning into a neoliberal global order. Indeed, globalizing neoliberalization proved to be a barrier for the implementation of a new phase of state-led hydraulic restructuring as much as increasing domestic dispute and contestation was. The old recipes no longer worked while new ones were easier to talk about than to implement. In the midst of this turmoil, a radical shift in the techno-natural edifice of Spain's hydro-social landscape was in the making. Chapter 8 will explore the contested production of this new hydro-social matrix.

8 Mobilizing the Seas: Reassembling Hydro-Modernities

If we could ever competitively, at a cheap rate, get fresh water from saltwater, . . . [this] would be in the long-range interests of humanity which would really dwarf any other scientific accomplishment.
—President Kennedy's ninth news conference, the State Department Auditorium, 4 p.m., April 12, 1961

On the morning of Friday, March 11, 2004, three days before national elections, ten explosions occurred on four commuter trains in the Madrid area. The attack killed 191 people and injured 1,800. The incumbent Partido Popular, which had a lead in the opinion polls, immediately accused ETA, the Basque separatist movement, of masterminding the attack. For the next two days, the PP insisted on the ETA link despite the latter's denial of involvement and early indications of an al Qaeda-related act. Prime Minister José Maria Aznar had been a major pillar of the "coalition of the willing" in the second Iraq war, a move that had met with significant discontent in Spain and elsewhere. On Saturday, evidence increasingly pointed to an extremist Islamist group as the culprit of the bombing. The investigation later confirmed the involvement of a group of activists loosely connected to al Qaeda. The next day, on Sunday, March 13, José Luis Rodriguez Zapatero of the PSOE unexpectedly won the Spanish elections, a feat largely attributed to Aznar's inapt handling of the crisis. The new government immediately withdrew all remaining Spanish troops from Iraq and disengaged from a deeply unpopular bellicose adventure. In addition, Zapatero suspended the most controversial parts of the National Hydrological Plan that had been approved in 2001 by Aznar's conservative administration. Discontent centered on the plan to transfer large quantities of "surplus" water from the Ebro river basin. Just a month after Aznar had dedicated the foundation stone for the Ebro transfer on February 18, 2004, the incoming new socialist Minister of the Environment Cristina Narbona called the

National Hydrological Plan history (Downward and Taylor 2007, 278). A new policy, appropriately called Programa A.G.U.A.,[1] formally approved on June 22, 2005, replaced the highly contested major river diversion schemes, and in particular the Ebro transfer, with a new hydro-technical logic, centered on the construction of twenty-one high-volume desalination plants on the Mediterranean coast—in addition to proposing measures for water recycling and reuse, and for improved water demand management—as the means to manage Spain's recurrent and endemic "water crisis." The desalination of seawater had indeed become one of the contested terrains for managing hydro-scarcities, particularly in arid and semi-arid regions.

Chapter 7 explored the contradictory process that had begun to erode the state-centered hydro-structural development that dominated much of the twentieth century while path-dependent power geometries attempted nonetheless to widen and deepen the prospects for a nationally integrated hydro-technical management of the terrestrial hydro-social cycle. This chapter examines the heterogeneous and often conflicting assembling of regional interests, political positions, seawater materialities, sustainability and ecological issues, sociocultural conflicts, and the demands of a neoliberalizing globalizing economy around desalination as a new socio-ecological and techno-natural "fix" for Spain. It considers how desalination and the networks of actors sustaining its realization mark the transition from a state hydro-structural framework to a decentralized, yet still very much state-led, market environmentalist water framework. I explore the socio-natural actors that assemble around this new, arguably revolutionary, socio-technological paradigm. The focus here is on the socio-environmental and ecological arguments advanced and how they gel in a new, but fragile, incoherent, and often contradictory assemblage that staged large-scale desalination as the emblematic discursive and material vehicle through which tension and conflict could be mediated and managed. Together with the arguments advanced in chapter 7, this chapter chronicles the tentative disassembling of the networks that sustained the hydro-structural hegemony and the embryonic reassembling of actants around desalination as new paradigm. I conclude that this assemblage of socio-natural actors enrolled around the desalination "fix" and the "mobilization of the seas" continues to focus on water supply (rather than other possible forms of socio-hydraulic management) and reproduce a consensual hydro-modernist development imaginary despite affirmations of radical change. Irrespective of the opposing arguments, both sociotechnical arrangements reproduce the existing hydro-development model, albeit with a reconfigured matrix of elite alliances and the mobilization of a new techno-natural dispositive, rather than signaling a radical new departure. In other words, this chapter tells the story about how sociotechnical configurations can change radically so that "nothing really has to change."

From Transfers to Desal: A New Hydro-Scalar "Fix"

El mar, fuente inagotable de vida.[2]

Spain was one of the first countries to experiment with desalination as a means for providing clean water. Even before the adoption of desalination as a large-scale and state-led alternative, a sociotechnical panacea for the country's water problem, Spain was already the fourth producer of desalted water in the world. In 2005, the country had a production capacity of 1.5 million cubic meters per day and between 700 and 900 mainly small and privately owned plants, enough to supply a population of eight million (Schmidt 2010, 32). The first desalination plant opened in 1964 on the volcanic Balearic island of Lanzarote (Meerganz von Medeazza 2008). Extending mainland terrestrial management of the hydro-social cycle into the sea as a "fix" for the conundrum of the country's uneven geographical distribution of water and for satisfying its unquenchable thirst for more water to fuel its development had been contemplated since the dying days of fascism. In 1973, for example, the *Revista de Obras Públicas* already insisted on the promises of mobilizing the seas:

> We should not see the future with pessimism, certainly not from Spain's perspective, so abundant in coasts, because, after all, the great alternative will be the sea. The sea will be our greatest reserve resource and we have it in abundance; . . . its exploitation is still rudimentary; . . . for us, it is the great unknown. If the means and the enormous human and economic efforts dedicated to the exploration of the stars . . . had been used in learning about the sea and obtain its benefits, we believe the results would have been far more productive. . . . From the sea we can expect great support of all kinds, which are now barely glimpsed."[3] (Valdés 1973, 409–410)

Recall how Costa (see chapter 3) at the turn of the previous century likened the hydraulic conquest to a military crusade, while—almost a century later—the desalination fix, in classic techno-culture hype, is pictured as a humanist-technological modernization effort comparable to presenting space exploration as the emblematic example of enlightened progress. While the regenerationist discourse eagerly mobilized a rallying military language, often filled with pathos and nationalist fervor, drenched in at times poetic, occasionally heroic, but invariably emotionally charged tones, the new post-democratizing rhetoric of water was couched in a much more technocratic, managerial, ecological and generally "scientific" jargon, only to be deflected occasionally when regionalist identity and nationalist demands became enrolled in the attempts to re-hegemonize the water discourse along new lines.

For the advocates of "desal" (desalination of sea or brackish water), the sea constitutes a seemingly endless, free, and uncontested source of unlimited supplies of water (March Corbella and Saurí 2008) that, with the right techno-managerial and political-economic support, can be combined with land-based water in a hydro-social and techno-natural assemblage capable of assuring the necessary water supplies. While terrestrial waters are marred by complex property rights, inserted in dense regulatory, institutional, and other legal arrangements, subject to all manner of social, cultural, and ecological conflict, and an integral part of often intractable multiscalar geopolitical inter-regional tensions and rivalries, seawater is seemingly free of these highly charged and conflicting meanings, contentious practices, and diverse claims. As such, the incorporation of the sea into the politics of produced water can be usefully identified as a scalar "fix" (Smith 1984) that can allegedly contain conflict by displacing the terrain of water mobilization to uncontested oceans while extending the scalar configuration of terrestrial hydro-social management by pushing the water frontier into the oceans.

As happened previously in Spain's history, water would once again become the emblematic "thing" around which much of the struggle and confrontation through which this transformation was enacted would crystallize. The discursive-cultural enrolment of the sea would became a kind of neutral space that permitted contradictory and conflicting arguments to coalesce around this strange attractor, the saline waters of the Mediterranean, that were staged as the panacea for Spain's water ills, yet allowed the continuation of a modernization-based growth trajectory articulated around expanding incorporation of nature within the circuits of capital accumulation.

The material qualities of seawater and the production of desal water were hailed together with the social, economic, and political merits of taking the water issue off the mainland. The production of clean water from the sea, so the argument went, would improve the overall quality of water as the purified and subsequently remineralized seawater would be mixed with lower-quality river or aquifer water, thereby improving significantly the overall physico-chemical quality of the water supplied to both households and industry (March Corbella and Saurí 2008). Obtaining permits to extract river or aquifer waters invariably met with a complex political and regulatory process. Combined with occasionally ambiguous property arrangements, the sea stood performatively for the site that offered unlimited and free—in the double sense of free of extraction costs and free of regulatory controls—access. High-technology reverse-osmosis desalination technologies were staged as the eco-innovation frontier for the global "sustainability" industrial complex in which Spanish companies took a leading edge. Moreover, such a decentralized and private industry-based water production process permitted adding water much more directly into a full economic accounting procedure and, thus, into a private profit-making industry articulated around the total commodification of water. The resulting higher price for water, particularly if desal was going to be used for irrigation purposes, was welcomed by some modernizing ecologists as a tactic to induce water-saving measures. Desalination would also, so the claim went, mitigate the unsustainable and often illegal and undocumented pumping of groundwater (Programa A.G.U.A. 2007). Furthermore, some autonomous regions, like Catalonia, staged decentralized desalination-based solutions as a sociotechnical configuration that permitted greater local control over resource management (and hence development) compared with large-scale and centrally orchestrated water transfers. Finally, the interminable conflict over alternative uses of the limited terrestrial water could be managed and contained by water from the sea (March Corbella and Saurí 2008). While inter-basin water transfers were very much part of an imaginary that linked it to authoritarian, top-down and bureaucratic politics, desalination was staged as "local," democratic, decentralized, private, market efficient, and ecologically sustainable. Dams and transfers were seen as archaic; desal was presented as high-tech, modern, and environmentally friendly.

Desal also became very much a crystallization point for choreographing identity and conflict between the leading political parties. The *Partido Popular*, with its strong popular support in the regions of the Levant, unequivocally supported water transfers. The Socialist Party tried to reinvent itself and draw on its original successes of fourteen years in power, which ended with José Aznar's electoral victory of 1996 and a subsequent eight years in opposition, by embracing a new vision for Spain's water development conundrum. While Aznar would basically continue the hydraulic politics of his socialist predecessor, the Socialist Party, eager to cash in electorally on the rising discontent against the water transfers, began to campaign for the environmentally modernizing, technologically innovative, and decentralized projects associated with desalination. Opposition to Spain's participation in the Iraq War and to the Ebro transfer became leading agendas in Zapatero's strategy to unseat the *Partido Popular*. The A.G.U.A. program of 2004 would cement this techno-natural transformation in a new policy framework.

The 2004 A.G.U.A. Program: "*Agua para siempre*" (Water forever)

Spain desalinates the gold of its coasts.[4]
—G. Schmidt, "España Desala el Oro de sus Costas"

In his investiture speech of April 2004, Zapatero fulfilled his electoral promise by announcing "a new politics of water, a politics that will take into consideration the economic value as well as the social and environmental value of water, with the objective to assure its availability and its quality, thereby optimizing its use and restoring the associated ecosystems"[5] (Zapatero 2006, 264). This rather generic statement announced a dramatic and radical shift in Spain's venerated hydraulic politics and signaled an emblematic moment, the pinnacle of more than twenty years of intense socio-environmental conflict, political contention, and proliferating discontent. The highly contested Ebro transfer enshrined in the 2001 NHP was suspended. As an alternative, the new A.G.U.A. program planned a total of twenty-one new or upgraded desalination plants along the Mediterranean coast with a total additional water production capacity of 1063 hm^3 per year, almost exactly the same volume as that promised by the Ebro transfer (see table 8.1). In an update and review in 2007, the Department of the Environment included two more desalination plants within the remit of the A.G.U.A. program (see table 8.2). Its full implementation would make Spain the Western country with the largest desalination capacity.

AcuaMed, shorthand for *Aguas de las Cuencas Mediterráneas, S.A.* (Waters of the Mediterranean Basins Ltd.), is a fully state-owned company that operates as a market entity. It is the main instrument of the Ministry of the Environment, Rural and Marine Affairs for the implementation of the A.G.U.A. program in the Mediterranean river basins. Its mission is the contracting, construction, acquisition, and exploitation of all hydraulic works of "public interest" that are undertaken in the basins of the Segura, Júcar, and Ebro rivers and in the Mediterranean basins of Andalusia and Catalonia, basically covering all water works from Girona in the North to Málaga in the South. The final objective is to produce an annual volume of desalinated water to the tune of 850 hm^3 per year. AcuaMed's commitment to desalination is unequivocal: "Our most valuable ally for obtaining water where there is scarcity, is desalination, the most economical, flexible and clean alternative to guarantee its supply regardless of meteorological conditions."[6] The total estimated investment cost was about four billion euro (approximately $5.2 billion) and several desal operations have already been implemented (see figure 8.1).

Table 8.1
Programa A.G.U.A.: Additional new water capacity, estimated cost, and planned desalination facilities

Province	Total additional capacity (hm³/year)	Total estimated cost* (Million €–2005)	New capacity, mainly desalination: New + upgraded (hm³/year)	Desalination plants	Cost of new capacity (Million €–2005)	Demand management (capacity gain) (hm³/year)
Girona	10	47	10	1	25	0
Barcelona	135	848	60	1	176	75
Tarragona	0	215	0	0	0	0
Castellón	78	173	33	2**	94**	32
Valencia	110	428	0	0	0	107
Alicante	212	618	141	7	292	71
Murcia	204	876	140	5	402	64
Almeria	189	352	165	5	226	24
Málaga	125	227	50	2	70	75
Albacete	0	14	0	0	0	8
TOTAL	1,063	3,798	599	23	1,285	456

Source: Ministerio de Agricultura, Alimentación y Medio Ambiente. 2011.

*Total estimated cost combines three measures: additional capacity production, demand management, and environmental improvement and flood protection.

**Added in 2006 and 2007 to the A.G.U.A. program.

Table 8.2
Desalination capacity—2007 (hm³/year). Total desalinated water capacity at the beginning of 2004: 140

Province	Operational 2007	Under construction	Adjudicated or in tender	Public inquiry	Planned	TOTAL
Girona	0	0	0	0	10	10
Barcelona	0	0	60	0	0	60
Castellón	0	0	18	15	0	33
Baleares	0	17	0	0	0	17
Valencia	0	0	8	0	0	8
Murcia-Alicante	54	219	31	0	30	334
Almería	42	20	50	5	0	117
Málaga	80	0	0	20	0	100
Canarias	0	9	10	0	0	19
Ceuta and Melilla	0	15	0	0	0	15
Total	176	280	177	40	40	713
Total cost (million €)	375	683	632	100	155	1,945

Source: Situación de las Plantas Desalinizadoras del Programa A.G.U.A.—Marzo de 2007, Ministerio de Medio Ambiente, Programa A.G.U.A., Madrid.

Figure 8.1
Desalination plant in Alicante.
Source: Author's photo.

This apparently fundamental departure from proven sociotechnical trajectories of supplying clean water unfolded through the disassembling of the networks of interests that gave shape and content to the hydro-structural configuration and the reassembling of different, occasionally overlapping, groups and interests around the desalination fix. In the next section, we turn to the heterogeneous collectives that were knit together through this reassembling process.

Assembling Heterogeneous *Actants*: Constructing the Desalination Edifice

An extraordinary reconfiguration of the networks of power has unfolded over the past few decades. In the process, traditional imaginaries and practices of hydro-social organization were radically overhauled, while nonetheless nurturing even further the insatiable appetite for the mobilization of additional usable water. The "symbolic universe" in which water dwelled eroded away as competing symbolizations attempted to forge gradually a new hegemony (del Moral Ituarte 2005, 52). While these new myths were often contradictory and antagonistic, a more or less stable, yet deeply heterogeneous and by no means hegemonic, assemblage emerged that proved nevertheless to be strong enough to turn the desalination discourse into real and new hydro-technical interventions. Four interrelated processes mark the passage to this new "symbolic universe." First, the emergence of a series of new mythologies around nature, novel ways of elevating nature to a matter of concern, which revisited what nature is, how it acts, and how humans can and should respond to it. Second, the relentless drive, both locally and globally, to water *"mercantilización,"* a process that combines the commodification and privatization of biopolitical life (Bakker 2010). While decentralized desalination facilities permitted full commodification of water supply, high-technology desal methods and the competitive advantage Spanish companies would obtain from nurturing its domestic desal development became actively staged as integral to Spain's competitive internationalization strategy. Third, the devolution of national state power augmented the hydro-social powers of local and regional assemblies in a context of intensifying inter-regional struggles and demands for regional autonomy. And fourth, the scalar transformation of the geopolitical and geoeconomic relations during the final quarter of the twentieth century, propelled the European Union as the preeminent geographical scale for environmental governance and political-economic regulation on the one hand and global neoliberal economic integration as the vector for economic growth on the other (see del Moral Ituarte 2005). The latter two processes are mutually constituted and define a process of what I have called elsewhere "glocalization" (Swyngedouw 1997). Ultimately, the manner in which desalination would herald "the end of water scarcity" would influence how these heterogeneous processes, arguments, and positions came together. Desalting the seas, so the argument went, would render water scarcity a thing of the past. Overcoming "scarcity," in turn, would permit massaging and managing the variegated but invariably escalating demands for water. As Hug March and colleagues put it, water desalination became "the new cornucopia" for Mediterranean Spain (March, Saurí, and Rico-Amorós 2014).

Water's Reimagined Acting

Sublime Aridity and the Management of the Hydro-Social Cycle
Matter matters politically, Jean Bennett argues in *Vibrant Matter* (Bennett 2010). Throughout this book, we have shown how indeed H_2O is politically active and significant, enrolled in specific ways in all manner of

discursive, material, technical, cultural, political, and social assemblages. The matter of water and the way it relationally articulates with other actors and *actants* shape the hydro-social edifice. This "acting" of water began to be reimagined in important new ways from the late 1970s onward, radically challenging dominant discourses and practices that had sustained water's role in specific techno-natural assemblages during most of the twentieth century. Whereas earlier discourses had considered "aridity" and water scarcity as nature's defect, a hostile and revengeful acting on the part of Nature against the Spanish People, on a par with the "aridity" of the sociocultural environment, recent reimaginations of water began to reclaim, both aesthetically and ecologically, the sublime landscapes of arid Spain. Through this rediscovery of the value of aridity, local waters became viewed as vital to maintain healthy arid ecosystems and to preserve their ecological value. While earlier, in the twentieth century, peasants fled from the pervasive and enduring doldrums of arid life, their urbanized children and grandchildren rediscovered aesthetically the "lost" but sublime landscapes of their forbears' lands (Lopez-Gunn 2009).

With the seminal ecological work of González Bernáldez, the arid landscape was indeed rescripted as containing a rich and drought-resistant plant and animal biodiversity of high aesthetic and ecological value (González Bernáldez 1981, 1989). This spurred a growing consideration of the effects of hydraulic infrastructures on arid ecologies and the importance of their hydro-social management. The semantic identification of aridity with "faulty," "defective," "poor," "infertile," "deficit," and "worthless" was replaced with a metonymic enchainment of signifiers like "scarce," "beautiful," "valuable," "rich," and "sublime." A similar semantic shift accompanied the growing concern with the fate of Spain's great wetlands. Once considered as insalubrious—mosquito and malaria infected—areas that required reclamation (Caprotti and Kaïka 2008) (like the Doñana area, Europe's largest wetland but now rapidly drying out), these wetlands too became viewed as cherished and crucial ecosystems increasingly protected under national and international law (García Novo, Toja Santillana, and Granado-Lorencio 2010). The reimagining of ecological relations and the central role of the now thoroughly engineered hydro-social cycle therein shifted the modalities through which water's acting became enrolled in the hydro-technical edifice. This revisioning of the value of both arid and humid ecosystems affected the vocabulary of "surplus" and "deficit" waters that had hegemonized most of the hydraulic discourse during the twentieth century. The preservation and nurturing of these "new" natures required a careful hydraulic management whereby linguistic signifiers like "scarcity" and "surplus/deficit" were no longer as politically performative as they once were. The major ecological challenge became the definition and management of the "right" volume of water and the implementation of ecological target conditions, adequate for the preservation of all water bodies and their associated ecosystems. As most water environments had been heavily modified, the emphasis was now shifting to the restoration of their "good ecological status." No longer were ecologies defined in terms of "dry" and "wet" but in terms of the correct and necessary amount of water to sustain their specific localized ecosystems and associated ecoservices. Soon the ecoservice language would of course also enter the neoliberalizing adagio of commodifying anything that might be constructed as "a service" under the ideologically dubious heading of ecoservice payments (Gómez-Baggethun et al. 2010). Achieving this required a transformation in the prevailing mobilization of Spain's waters that hinged on watering the "infertile" dry lands that "lacked" the necessary water to be useful, productive, and beautiful. However, the preservation of "basic ecological water flows" (Manteiga and Olmeda 1992; Palau 2003) hit the escalating water demands of the usual suspects: irrigators and land developers. Here too, the waters of the sea became mobilized as the terrain that could satisfy both demands: maintain and sustain terrestrial water flows to assure good ecological status, equitable access, and expanding supply.

With a Little Help from Climate Change ... and Malthus
Over the past decade or so, the global environmental conundrum focused on climate change as the material and symbolic condition around which our socio-ecological predicament circulates (Swyngedouw 2010a, 2011). Here too, the concern of both economic elites and ecological modernizers alike focuses on how to sustain capital accumulation and economic growth for a while longer, while mitigating or adapting to rapidly changing socio-environmental conditions that are often portrayed as potentially cataclysmic. This is particularly acute, as climate change has now entered multiple discursive formations, whereby economic actors, policymakers, scientists, as well as broad strata of civil society and activists vie for climate-concern respectability.

A curious discursive fusion of the actions of salty water and of CO_2 has hegemonized part of the desalination argument in Spain, whereby ecological concerns and arguments are mobilized to argue for an increase in water production that is secure "in guaranteeing water, rain or no rain, independent of the climate" (ICEX 2010, 3). The latter's reference to "external" meteorological conditions—repeated ad nauseam by AcuaMed representatives—is of course a stand-in for "climate change," which remains as such unnamed. Yet, the debate over future water availability and supply is invariably marred by the specter of climate change. The Fourth Assessment Report of the Intergovernmental Panel on Climate Change (IPCC) (Pachauri and Reisinger 2007) does indeed paint a rather bleak picture of the Mediterranean's water futures and likely changes in the hydrological cycle. Climate change projections indicate a decrease of precipitation in the already highly water-stressed southern regions of up to 40 percent by midcentury compared to average 1961–1990 levels, and a small increase in the northern regions. Data from the Department of Agriculture and the Environment suggest a noticeable reduction over the past decade in the average annual flow of all river basins, but particularly those in the southern regions (see table 8.3). Ecologists invariably invoke the specter of climate change as the main cause of this significant decline. Using meteorological data that suggest an increase of the average temperature by 0.48 °C during the 1973–2005 period, the resulting rise in evaporation is cited as the principal driver of declining river flows (Martín Barajas 2010, 61).

The IPCC's expected overall and now largely inevitable temperature increase of at least 1.5°C this century will indeed lead to higher evaporation, reduced soil moisture, and increased crop evapotranspiration. The overall predicted effect on reservoir water availability is estimated to be around –5

Table 8.3
Average annual flow in the main rivers, 1940–1955 and 1996–2005 (hm^3/year)

River basin	Average flow		Percentage change
	Period 1940–1955	Period 1996–2005	
North	43,494	38,573	–11.3
Duero	13,861	11,729	–15.4
Tajo	10,533	9,012	–14.4
Guadiana	5,464	4,391	–19.6
Guadalquivir	8,770	8,113	–7.5
Mediterranean Andalusia	2,446	2,101	–14.1
Segura	817	505	–38.2
Júcar	3,493	3,057	–12.5
Ebro	17,189	13,555	–21.1
Internal to Cataluña	2,742	2,196	–19.9
Total	**108,809**	**93,232**	**–14.3**

Source: Department of the Environment, Agriculture and the Sea, Madrid.

percent to -7 percent. The Centro de Estudios y Experimentación de Obras Públicas (2011) predicts a generalized reduction of precipitation of -5 percent between for the period 2011–204, rising to -9 percent between 2041 and 2070, and a whopping -17 percent between 2071 and 2100. The greatest variability will take place along the Mediterranean coast and in the southeast. The predicted temperature rise will, furthermore, increase evaporation and evapotranspiration, and decrease groundwater recharge and runoff (Vargas-AmelinPindado 2013).

The European Environment Agency's assessment of the sociohydrological conditions as a result of climate change for the Mediterranean areas predicts an increasing demand for agricultural water, increased summer water shortages, and deteriorating water quality due to higher water temperatures (European Environment Agency 2007). The specter of climate change is then hailed as one of the contextual driving forces that point to desalination as a possible alternative water supply technology to mitigate the consequences of climate change on the hydro-social cycle and to facilitate adaptation to the new socio-physical environment. On the basis of such compelling data, an extraordinary consensus has emerged among business leaders, elite institutions, and some environmentalists that the twin forces of climate change and demographic expansion point to desalination as potential sociotechnical "fix":

> Desalination is increasingly being used to provide drinking-water under conditions of freshwater scarcity. Water scarcity is estimated to affect one in three people on every continent of the globe, and almost one fifth of the world's population live in areas where water is physically scarce. This situation is expected to worsen as competing needs for water intensify along with population growth, urbanization, climate change impacts and increases in household and industrial uses. (World Health Organization 2011, 1)

This view is widely shared by a host of national and international organizations (see, for example, Boyé 2008) and legitimized by a growing number of scientists. Menachem Elimelech and William Phillip, for example, argue in a paper published in *Science* that "population growth, industrialization, contamination of available freshwater sources and climate" constitute key challenges for water provision. It is important to find a way to alleviate this stress with a sustainable solution and point to large-scale seawater desalination as one of the few options available (Elimelech and Phillip 2011 712).

Global hydro-technical corporations are quick to pick up and invoke the twin pressures of the Malthusian specter and climate change adaptation to advertise and advocate the pursuit of innovative and "sustainable" desalination projects. The American Water company, for example, in a White Paper on desalination notes:

> The world's population is expected to increase by 50 per cent by 2050, but only one per cent of the earth's water is freshwater that is ready for drinking, and the number of regions experiencing droughts and water shortages is growing. . . . With 97 per cent of the earth's water consisting of seawater, one viable solution is desalination. Desalination. . . has been successfully implemented around the world and has proven to meet the needs of residents that would otherwise have no local access to drinking water. Though not widely used at a large scale in the U.S., desalination is beginning to make headway across the country, particularly in arid coastal regions. (Duffy n.d., 1)

Siemens, the German giant, asserts that "water scarcity, due to climate change and population growth, affects one-third of the world's population. Siemens can help you to augment your water resources—safely and reliably—with innovative solutions for water reuse and desalination."[7] Spanish companies too emphasize the magical powers of desalting the seas. Acciona, partner of American Water in Florida's Tampa Bay desal project, inaugurated in 2010 London's first large desalination facility on the Thames River at Beckton. Commissioned by Thames Water—London's water supplier that is partly owned by the Australian investment fund Macquarie, the China Investment Corporation, the Abu Dhabi Investment Authority, and the British Telecom Pension Scheme—the plant is designed to supply water irrespective of climatic conditions. As Acciona asserts:

> London is a city affected by hydric stress, and without adequate planning a year of low rainfall could give rise to severe water shortages. London's average annual rainfall is lower than that of hotter cities such as Istanbul, Sydney or Dallas. London's population is growing and climate change is threatening the UK capital with hotter, dryer summers in the future. The Beckton plant falls within the *Gateway* Project, a long-term plan designed by the City authorities and which sets out to avoid water shortages in London, even in times of drought.[8]

Befesa,[9] a major Spanish desal company, joins the chorus, as does Veolia, one of the world's leading water companies and contractor of the Sydney desalination plant: "The desalination plant will ensure Sydney has a secure and reliable supply of water that is not rainfall dependent. This contract comes in the face of uncertainty over climate change and strong demographic growth in the region."[10]

Abeima, another Spanish player in the rapidly expanding market, insists boldly that "desalination is an alternative and inexhaustible source of water at a moment when the supply of this resource has become a real problem in many place of the world, aggravated by the condition of climate change."[11]

Indeed, the global water industry is rising up to the challenge of a warmer world and the opportunities for accumulation by desalination this offers. How this articulates with the Spain's environmental modernization drive will be discussed in the next section.

"Mercantilización": Marketing Water

Liberal Green Managers to Save the Planet (and Neoliberal Spain)

Growing environmental concern fused almost seamlessly with a new managerial-technocratic and economic rationality. The rescripting of how nature and water act and in what ways this matters paralleled the widespread resistance to building additional large terrestrial hydro-technical infrastructures. The political-economic transformation of post-dictatorship Spain to a neoliberal state and its corresponding rationality began to define the water problem in new terms. Who pays, who benefits, what are the economic and ecological values of water projects, and issues of equity, democracy, and efficiency became key leading concerns that began to suture the water debate. These debates engaged a wide and heterogeneous coalition of activists, academics, and water administrators that gathered and organized to demand a "New Water Culture"[12] (NWC) (Martínez Gil 1997; Arrojo Agudo 2001b, 2005; Aguilera Klink 2008). Ecological concern, sustainable development, rational water management, demand-side regulation, and local empowerment were the main axes along which the demands for a new water culture became articulated (Tabara and Ilhan 2008). The "engineers of steel" (Abad Muñoz 2006, 47), an acerbic nickname for the hydro-technical architects of a bygone era, became identified with a centralized and authoritarian state apparatus, unlimited cheap and subsidized water, technocratic rule, a disdain for "real" nature, and blinkered views of what constituted "development." Instead, NWC advocated decentralized and participatory integrated management, ecological sensitivity, the mobilization of multidisciplinary perspectives, and, above all, a rational valuation of the economic, social, and ecological value of new projects and policies. "Water as life" became the guiding biopolitical principle, replacing the engineering dictum of water as a mere means of production (Lopez-Gunn 2009, 378). When the PSOE announced the *Programa A.G.U.A.* in 2004, Cristina Narbona Ruiz, the newly appointed minister of the environment, embraced the principles of the NWC as a major departure from the past and expressed her commitment to its principles, which—she claimed—were an integral part of the motivation that spurred the newly adopted desalination policy. Full cost recovery, decentralized organizations, and the mobilization of market principles, but embedded in a strong eco-regulatory framework, replaced the old markers of the hydro-structural paradigm. While the traditional hydraulic engineers lamented the fading prestige (Abad Muñoz 2006, 47) of their terrestrial technocratic solutions, environmental modernizers embraced the new challenge. Demand management, "real" pricing of water (to reduce inefficient water use in agriculture), and the mobilization of technological fixes for environmental purposes began to hegemonize hydro-social discourses. In fact, hydro-environmental modernization quickly turned into a new fix that would save the environment while opening a new space for capital accumulation.

Environmental Modernization and Globalization: Accumulation by Desalination

We project that by 2014, we will be adding more than the equivalent of a new River Thames each year to the world's renewable freshwater resources. By 2020, the seawater desalination industry will be adding twice that amount. We are creating rivers that flow backwards from the sea. . . . We forecast that [seawater desalination] will be a $16bn industry in 2020.[13]

—Christopher Gasson, Global Water Intelligence, August 9, 2009

The Spanish Institute for Foreign Trade (ICEX), within the Department of Industry, Tourism and Trade, identified water desalination as one among eight activities that showcase Spanish technological innovation and excellence at a global scale. ICEX deliberately and explicitly stages desal as an innovative and economically promising alternative to the outdated, outmoded, and plainly "ineffective" water transfers (ICEX 2010). Brandishing the environmental banner and mobilizing the financial and political clout of the Spanish state to promote desalination through the A.G.U.A. program, the Spanish desalination industry, with its own lobbying organization (The Spanish Desalination and Water Reuse Association, AEDyR), takes its successes in the national market as a springboard for international expansion and for carving out a lucrative and competitive edge among the leading global desalination players in a rapidly expanding global ecomarket that sees itself at the forefront of combating water scarcity in an environmentally sustainable and climate-friendly manner (ICEX 2009, 2010). "The announcement of the plans to develop these new desalination plants has been a boon for desalination companies" (ICEX 2010, 3), the trade promotion literature states. Spanish business and political elites clearly manage the collective effort to change the techno-natural configuration as a preemptive gesture geared at solidifying Spain's high-tech industry's competitive position in the global marketplace. ICEX, for example, states how "at times the companies are competitors when submitting bids for new plants. . . . At times the companies work in various consortia. . . . The Spanish government, in an effort to support a variety of Spanish companies, divided the development of the landmark Carboneras project [among some of them]" (ICEX 2010, 4). Collaborating at home is staged as a launching pad for Spain's hydraulic expertise to gain access to foreign markets and for improved global competitiveness. As José Antonio Medina, president of AEDyR, put it: "We have been working for the past 30 years on all these desalination plants. That gave the Spanish companies the necessary expertise with both building and operating plants. At the moment, Spain has the highest number of companies in the world with this level of technology and experience in desalination" (Medina, cited in ICEX 2010, 3–4).

Desal is indeed an exponentially growing business. In 2010, more than 65 million m^3 per day of water is desalted globally in more than 14,000 installations, up from 5 million only thirty years ago (see figure 8.2). More

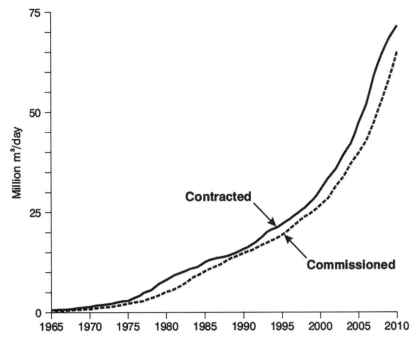

Figure 8.2
Global cumulative contracted and commissioned desalination capacity, 1965–2010.
Source: Global Water Intelligence—IDA Desalination Yearbook 2010–2011, http://www.desalyearbook.com/.

important, the new Spanish desalination plants all use reverse osmosis (membrane) technology, which is considered to be the most efficient, leading-edge, and promising high-tech solution, compared with the expensive and energy-intensive old-fashioned thermal or hybrid technologies (see figure 8.3). Spain is now placed fourth in the world (after Saudi Arabia, the United States, and the United Arab Emirates [UAE]) in terms of total installed capacity. However, it is catching up quickly as it has jumped to third place (before the United States) in terms of newly installed capacity since 2005 (see figures 8.4 and 8.5) (Global Water Intelligence 2010). No less than eight Spanish or mainly Spanish companies figure in the top twenty of the world's leading desalination contractors by volume of desalinated water. Seven of the leading reverse osmosis membrane producers are Spanish too (see figures 8.6 and 8.7). Accumulation by desalination has indeed been a success story. The Spanish state-industry partnership is playing a lead role in this. As an ICEX publication, aimed at the U.S. market, gleefully notes: "Spanish firms' know-how, flexibility and expertise allow them to be

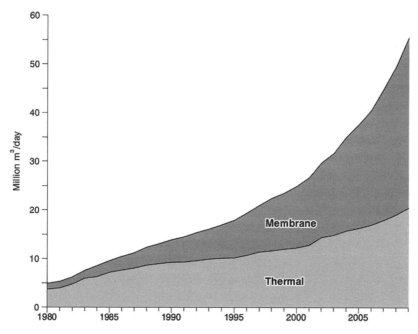

Figure 8.3
Installed membrane and thermal desalination capacity, 1980–2009 (cumulative).
Source: Global Water Intelligence—IDA Desalination Yearbook 2010–2011, http://www.desalyearbook.com/.

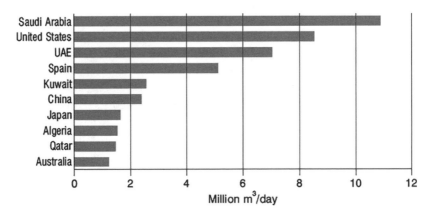

Figure 8.4
Top ten countries by total installed desalination capacity since 1945.
Source: Global Water Intelligence—IDA Desalination Yearbook 2010–2011, http://www.desalyearbook.com/.

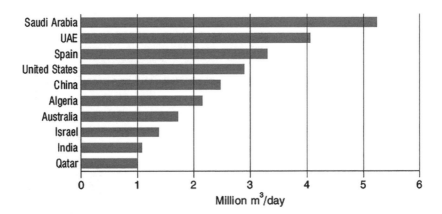

Figure 8.5
Top ten countries by total installed desalination capacity between 2003 and 2010.
Source: Global Water Intelligence—IDA Desalination Yearbook 2010–2011, http://www.desalyearbook.com/.

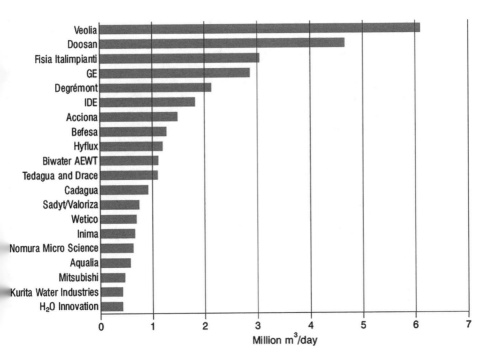

Figure 8.6
Top twenty engineering, procurement, and construction (EPC) desalination contractors, 2000–2010 (by contracted capacity).
Source: Global Water Intelligence—IDA Desalination Yearbook 2010–2011, http://www.desalyearbook.com/.

successful even in the most competitive markets." (ICEX n.d., 2). All of this is of course the result of Spain's longstanding commitment to desalination, another ICEX publicity document aimed at the U.S. market notes: "Spain built Europe's first desalination plant nearly 40 years ago and is the largest user of desalination technology in the Western world. Spanish companies lead the market, operating in regions including India, the Middle East, and North America. Spanish innovation contributes to advancing desalination to bring sustainable clean water to millions of people" (ICEX 2010, 1).

And reverse osmosis technologies are thereby the wave of the future: "This long tradition and experience in the field has turned Spanish water-treatment and desalination companies into world leaders in the technologies involved, such as reverse osmosis. Many of them are subsidiaries of large construction companies, with the clout and financial acumen to undertake large turnkey projects all over the world, either by themselves

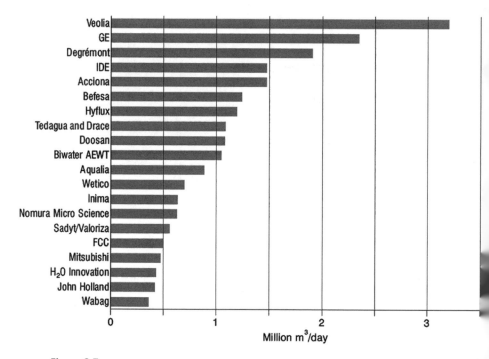

Figure 8.7
Top twenty membrane-technology desalination contractors, 2000–2010 (by contracted volume).
Source: Global Water Intelligence—IDA Desalination Yearbook 2010–2011, http://www.desalyearbook.com/.

or through consortia or joint-ventures" (ICEX n.d., 2). Figure 8.8 shows the map of global ventures in desalination operations in which Spanish contracting companies are actively involved. The success at home brought success internationally, with Spanish desal companies, as well as their supporting niche suppliers of high-tech components, now active operators on all continents in major and globally significant projects. It is indeed necessary, so an industry report notes, "to progress steadily in foreign markets like Saudi Arabia, Algeria, Australia, China, UAE, India, Libya, Morocco or Tunisia" (Fundación Cajamar 2009). In the United States, Spanish companies are actively involved in Tampa Bay, Florida (Pridesa), and in Brockton, Mass. (Inima). In Algeria alone, which is building seven large desalination plants, five will be constructed by Spanish companies and one by a Canadian-Spanish consortium. Only a few U.S. companies (such as General Electric Water and the Doosan Conglomerate) and the two French water giants (Suez and Veolia) play in the same league.

Mobilizing the Seas

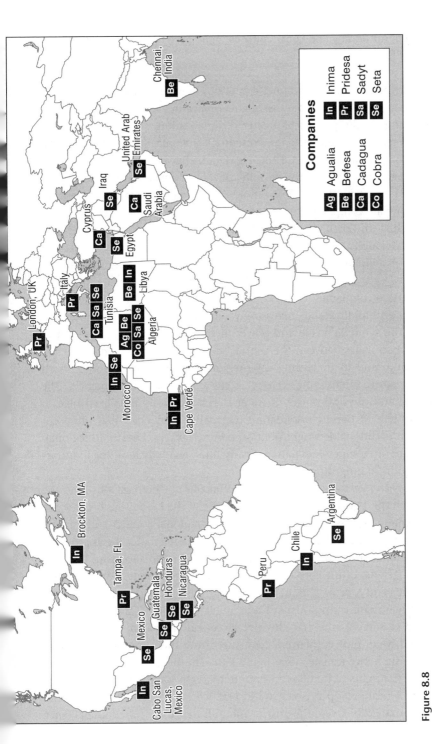

Figure 8.8
Major international desalination operations by Spanish companies.
Source: ICEX 2010, http://www.spainbusiness.com/icex/cma/contentTypes/common/records/mostrarDocumento/?doc=4119833 (accessed June 2, 2011).

Regional Autonomy and Local Cultures: *Agua para todos* versus a Knotted Water Pipe

The turn to the seas also implies a "politics of scale," a reconfiguration of the articulation of various scales of governance in a new "scalar gestalt" of governing. In the process of greater regionalization, several regions have created regional water agencies. Presently, Catalonia, Castilla-León, Andalucía, Aragón, Murcia, and Castilla-La Mancha have all established regional water agencies with partly or fully devolved control over their water affairs (Lopez-Gunn 2009, 384). In addition, state-organized—but operating as private companies—*sociedades estatales* (state companies) have been established (in 2010), after the 1999 amendment of the 1985 Water Law, as private-public partnerships to organize the planning and implementation of hydraulic works. AcuaMed is one of these societies; the others are AcuaNorte, AcuaSur, and AcuaEbro. They replace the earlier nine river basin–based companies.[14]

In the unfolding and increasingly complex scalar reconfiguration of Spain, the symbolic value and power of water is affectively mobilized in political struggles between the central state and regional identity on the one hand, and in fomenting inter-regional conflict on the other, both galvanized by intensifying regionalist cultural identity construction. The cultural and physical power of water is systematically mobilized here in the context of these consolidating regionalist tendencies that often cut through traditional left-right political party cleavages. Loyalty to regionalist claims often supersedes national party lines, regularly resulting in intractable political infighting.

The *guerra del agua* (water war) between the northern regions (and the waters of the Tajo and Ebro rivers) and those of the Levant has become the emblematic expression of intensifying inter-regional conflict around who has the right to water and for what uses. Immediately after the approval of the National Hydrological Plan, the highly active and successful Platform for the Defense of the Ebro River was established. The water donor regions, Catalonia and Aragón, and the receiving regions, Valencia and Murcia, took radically opposing positions. As exemplified in chapter 7, the Platform succeeded in staging an extraordinarily successful campaign of resistance. A knotted water pipe is their emblem, suggesting their opposition to inter-river basin water transfers (see figure 8.9). The fusion of ecological arguments, the rights of humans and nonhumans to water and defense of the intimate relationship between water practices and local/regional identity (Mairal Buil 2005) were the grist to the mill that forged the loose alliance

Figure 8.9
A "knotted" pipe: emblem of the Platform for the Defense of the Ebro.

of academics, green activists, and regionalists into a formidable opposition force (Masjuan et al. 2008). Of course, the receiving regions, Valencia and Murcia in particular, defended with similar zeal *Agua para Todos*, appropriating UNESCO's powerful "Water for All" slogan to make a case for "an equitable and just distribution" of Spain's water to make sure the development and identity of Murcia could be preserved and nurtured further. A crusade of local entrepreneurs, farmers, real estate developers, and an alliance of both right- and left-leaning parties in Valencia and Murcia supported and continued to defend the full implementation of the NHP, including the water transfers. A series of mass-demonstrations showed their determination to have the transfer project implemented. After the derogation of the Ebro transfer in 2004, the controversy continued and even intensified. In 2005, for example, the socialist president of the region Castilla-La Mancha, José María Barreda, stated: "we are engaged in a long water war that Castilla-La Mancha will win" (Mendez 2005). During the severe drought in Catalonia in 2008–2009, calls from Catalonia to transfer water from the Ebro to alleviate water shortages met with serious opposition from neighboring Aragón (Arrojo and Visa 2009). Further escalation of the "water war" touched upon the spinal cord of the national water system (the Tajo-Segura water transfer) when the government of Castilla-La Mancha announced that it would unilaterally end water transfer by 2015, thereby further exploding the finely tuned regional-national agreements that had been carefully negotiated over the past two decades. Supported also by the regional branch of the Partido Popular, the statement hit like a bombshell, immediately setting off a series of vitriolic responses from the receiving regions that considered the initiative to be a direct onslaught on their regional survival and a usurpation of power over water that belongs to the national state. A study by the University of Murcia and contracted by the PSOE estimated—in a medium-range scenario—that the end of the Tajo-Segura water transfer would entail the loss of 51,500 jobs and an economic loss of 1.7 billion euro (Sancho Portero 2008).

The water wars contributed decisively to reinforcing the identity of regions that hitherto had not explicitly politicized the regional question. In Murcia, for example, a new "hydraulic nationalism" emerged, particularly after the suspension of the Ebro-Segura transfer and the threat by the ceding regions to cancel the Tajo-Segura transfer. These new forms of regionalism intensified already acute inter-regional tensions, and added a further layer of complexity to the already intractable and conflict-ridden national-regional relations. As the Citizens' Forum of Murcia put it:

> In contrast to other autonomous regions, the Region of Murcia does not have a nationalist party, neither has it been endowed with an imaginary national identity, nor has it affirmed itself against other autonomous regions as a unitary and excluding unity. However, since 1995, it has been the PP [the conservative party is hegemonic in Murcia], the party that supposedly guards the national unity of Spain, which has managed to create out of nothing a patriotic sentiment of being *"murcianía."*
> ... But this new feeling is based on a peculiar hallmark of identity: water has been turned into a great common totem of the *"Murcianos"* by means of the simple and repeated slogan of "water for all."[15] (Foro Ciudadano 2005; also cited in del Moral Ituarte 2008)

This hydraulic regionalism transforms and competes with twentieth-century Spanish nationalism and is based, The Citizen's Forum argues, on the interweaving of three key arguments (Foro Ciudadano 2005). First, *"el victimismo"* (the nurturing of a culture of victimhood), blaming "the others" who "assault us and rob what is ours," in this case the Catalonians, the Aragonese, and the "Manchegos" because "they do not provide us with the water we demand and to which we believe we have a 'historical right.'" Secondly, a budding narcissism articulated around the view that Murcia is better than the others in using and managing water more effectively and efficiently: "the others just waste water and take it to the sea." Finally, the consensus that is articulated through the "Water for All" slogan around which the regional government party and opposition, entrepreneurs and unions, developers and farmers, savings banks, and the media rally, while those who disagree are treated as "traitors" and infiltrators operating in the service of "the others." This strong language suggests the extent of the passions that erupt around the regional water question.

While there is regional consensus on the right to water transfers, desalination is simultaneously celebrated as permitting a greater autonomy for the regions of the Levant and assuring their regional economic and cultural independence. The vice-president of the Polytechnic University of Cartagena summarizes the dominant view well, whereby the techno-social fix of desalting the seas offers a way out:

Fortunately, the region of Murcia has a coastline which allows an infinite availability of this primary resource. This condition guarantees the elimination of all the uncertainties or availability that necessitated the transfers, because the application of this technology [desal] permits the supply of water similar to those planned from the Ebro transfer. . . . However, the supply of resources from desalination cannot be isolated from other waters of the basin, in the sense that resources that previously came to the coastal areas from upstream [through the transfer] and that will now come from desalination, can and must be dedicated to supplying the high and intermediate zones of the basin. . . . In other words, desalination must be considered together with all the water available in the system.[16] (Castro Valdiva 2007, 8)

The "Murcianos" clearly want to have their cake and eat it. A more sober view of the escalating inter-regional conflicts around water as the state-region relations in Spain are reworked into a new scalar "gestalt" recognizes that the intensification over the territorial configuration of the hydro-social cycle renders national compromise and consensus ever more difficult to achieve. The sea surfaces here again as a "scalar" fix that permits attenuating these intractable geographical tensions. The daily *La Vanguardia* expressed the sentiment of many when it wrote about the "marine solution" that would attenuate and contain escalating conflicts around water: "from the sea will come the water that will avoid the territorial conflicts emerging during every drought period" (cited by March Corbella 2010, 349). In sum, the decentralized nature of desalination factories and the use of marine waters reinforce the regionalization of the hydro-social cycle and fuel further inter-regional conflict. The techno-natural configuration of desal permits indeed aligning the water conflict with the desire of some to strengthen regional autonomy from what are considered the invasive powers of centralized Spain in Madrid.

The European Union's Water Framework Directive and the Rescaling of Water Governance

While the Spanish state decentralized after the dictatorship, a major, parallel process of governance upscaling unfolded. In 1986, Spain joined the European Economic Community. "Becoming European" and joining European "modernization" was a shared objective of the democratization process after the dictatorship, "an emotionally charged symbol of national resurgence" that would propel Spain out of its backwardness, solidify democracy, and entrench modernizing desires of progress and development, whereby "national sentiments smoothly intertwined with the process of European Integration" (Jáuregui and Ruiz-Jiminéz 2005, 73). With the entrenchment of the transfer of significant policy domains to the European scale and the unequivocal adoption of neoliberal economic rule—although mediated through social and environmental trans-national regulation, starting with the Maastricht Treaty of 1992 and consolidated in the 2007 Lisbon Treaty——a complex multiscalar network of market-led technocratic governance emerged. Translated in socio-ecological terms, Europeanization coincided with a drive toward environmental modernization that articulated the commodification of nature with new forms of market-driven and rationally formed environmental governance (Heynen et al. 2007; Birch, Levidow, and Papaioannou 2010; Furlong 2010).

The European Water Framework Directive (WFD) enshrined these principles in the water sector, precisely at a time when the Spanish water transfer debate reached a new apogee. The European Water Framework Directive set a generic European Union-wide policy framework for national water policies (CEC 2000; Page and Kaika 2003). The main drivers of the framework combined a requirement to strive toward full cost recovery of water infrastructures, thereby eliminating structural or systemic subsidies to water users and nurturing rational and careful use of a scarce resource with a focus on maintaining and/or restoring "good" water quality, which included the protection of aquatic systems. Moreover, the WFD stipulated that the governance of water should be organized on the basis of river basin organizations that would also ensure inclusive participation of all relevant stakeholders (Kaika 2003). The WFD was approved in 2000 and translated into Spanish Law in 2003 (Grindlay et al. 2011), precisely during a period of intense conflict over and transformation of the hydraulic paradigm (Saurí and del Moral 2001). The complex and highly contested implementation process fused with existing socio-environmental conflicts and debates explored earlier, and resulted in a cacophony of competing and invariably highly charged political debates (Thiel, Sampedro, and Schröder 2011). The WFD was invoked to critique water transfers, state subsidies, hierarchical planning, and autocratic forms of governance, and to defend greater ecological sensitivity, market-led pricing, full cost recovery, and decentralized "participatory" governance. Each of these principles was successfully mobilized by those who opposed the large-scale water transfers foreseen in the National Hydrological Plan. The resulting intractability of deeply entrenched positions further eroded the increasingly precarious stability of the traditional hydraulic paradigm and hastened its transformation to managerial sociotechnical fixes able to contain and massage existing tensions and conflicts while maintaining a relatively coherent eco-modernizing development trajectory (Font and Subirats 2010). The marine "solution" offered precisely an opening to adopt the principles of the WFD while also opening an avenue for assuaging the demands of those who insisted on a greater supply of freshwater.

The Return of the Real and the Contradictions of Desalination: "How to Change Radically so that Nothing Has to Change"

Despite the political, economic, and ecological advantages it offers, particularly with respect to enabling Spain to continue as usual with its water-gulping growth-based trajectory, the heterogeneous socio-physical assemblage that sustains the desalination edifice is nonetheless a fragile one as internal tensions and contradictions destabilize and rupture the alliance that permitted the desalination fix to materialize. A permanent jockeying for position between the transfer and desal options continue to choreograph the dispute over the hydro-social configuration. For example, after the electoral defeat of the Socialist Party in the elections of 2011, the A.G.U.A. program—a cornerstone of José Zapatero's hydraulic vision—quietly disappeared from the political discourse of the successor conservative government headed by Mariano Rajoy. Even the dedicated website was removed.

The bones of contention and controversy revolve around five interrelated claims and arguments. First, ecologists and other activists are concerned about the effects of waste generated by large-scale desalination. Key among them are the highly saline residual waters that need to be released back into the environment on the one hand, and the rising CO_2 output as a result of the high energy demands of desalination on the other (Meerganz von Medeazza 2005). The brine is usually mixed with seawater and pumped back into the sea. Marine ecologists are concerned about the effects of the physico-chemical composition of the residual waters on marine ecosystems. Climate change specialists add to the ecological concern by pointing out how the energy demands (circa 5 KWh of electricity for one m^3 [Meerganz von Medeazza 2008]) of desal water accelerate the accumulation of CO_2 in the atmosphere as Spain's energy mix is now primarily fossil-fuel based. Second, the higher price for desalination water (circa 0.50 euro/m^3 off the plant) compared with transfer water (circa 0.2 euro/m^3) or pumping ground water (circa 0.3 euro/m^3) has sparked significant problems (Rico Amorós 2010). Both water distribution companies and irrigators are reluctant to pay for desalination water given the high price. While initially the price of desal water for irrigation was to be subsidized, the WFD specifically prohibits just such state support. Some of the new desalination plants in Alicante and Murcia are actually operating at reduced capacity (thereby further increasing cost per unit) because overall demand remains stable as the higher price for desalinated water encourages water-saving efforts and, most problematically, irrigators often resort to expanding illegal, but cheaper, groundwater pumping, thereby aggravating already precarious groundwater availability and quality. Third, the pumping of "gratis" seawater has sparked debate over legal rights over the seas and the "cost" of seawater as a resource. This will undoubtedly intensify debate over who has jurisdictional powers over coastal waters and it has already fuelled critical debate over the "privatization of the seas." Fourth, the multiscalar financial flows and support mechanisms that finance the desalination effort (the European Union, the national and regional governments, private companies) intensify debate over the mobilization of public resources to nurture specific interests and render the multilevel governance configuration extraordinarily complex and internally fractious. Finally, and perhaps most important, the desal controversies have opened up a new, significant wedge in the fragile hydro-social configuration.

While it is becoming increasingly clear that the desalination hydro-social fix inscribes itself in a reworked expansionary developmental logic in which water transfers combine with other modes of water management, a more radical critique has emerged in recent years that questions the hegemony of this hybrid mix that is staged as the panacea to manage the hydro-social cycle such that the agro-tourist-urban growth machine can be sustained into the foreseeable future. A fledgling de-growth movement mobilizes around the need to slow down or reduce economic growth in order to produce ecologically saner and socially more just conditions (Kallis 2011). Furthermore, rising general discontent against the speculative boom of the past decade, the undemocratic character of neoliberal governance, the socio-ecologically negative excesses of market-based growth, and the political marginalization of alternative voices have produced a powerful *indignado* (the outraged) movement that seeks a more radical and democratic sociopolitical transformation of the Spanish political-economic landscape. While they do not address water directly, the claims for real democracy and greater justice reverberate throughout Spanish society and provoke considerable concern among its elites. Indeed, despite the selective hailing of desalination as a techno-natural revolution capable of saving Spain from thirst, desal is increasingly seen as a socio-natural fix that permits a productivist water logic to remain the bedrock of Spain's modernization project so that "nothing really has to change" (di Lampedusa 1960).

9 Politicizing Water, Politicizing Natures: Or . . . "Water Does Not Exist!"

Let nothing be called natural
In an age of bloody confusion,
Ordered disorder, planned caprice,
And dehumanized humanity, lest all things
Be held unalterable!
—Bertolt Brecht, from the prologue to *The Exception and the Rule*

Beyond H$_2$O

The journey through more than a century of Spain's tumultuous historical and geographical transformation has come to an end. Throughout, I explored how the socio-ecological meanderings of water became etched into the transformation of the hydro-social cycle. This process of hydro-modernization laid bare the variegated relations of power through which water becomes enrolled, transformed, and distributed. I showed how the matter of nature enters the domain of the political and, through this, how environmental reconfiguration parallels the continuous socio-spatial transformation of both state forms and social orderings.

The story was wedged between two emblematic crisis moments. It began with *El Disastre*—the inglorious end of empire in 1898—and the painful convulsions of a country that found itself in a traumatic postimperial condition as the geopolitical coordinates of imperial constellations shifted rapidly; it ended, more than one hundred years later, in 2010, also a time deep economic crisis, widespread cultural anxiety, intensifying social conflict, and geopolitical transformation. The book presented Spain's turbulent twentieth century as a political-ecological project marked by profound changes, punctuated by periods of great hope and expectations, intense social and political conflict, democratic reform and political closure, brutal civil war and dictatorship, shattering crises and remarkable socio-spatial and cultural change.

The political-ecological processes that shaped these transformations were mobilized as the lens through which a wider set of issues related to nature, the environment, modernity and political power were explored. This book offered a story of the extraordinary meanderings of H_2O, but also aspired to grapple with water and the hydro-social circulation process as the hybrid fusion and heterogeneous assembling of things human and nonhuman. Indeed, the hydraulic techno-natural revolution of Spain's territorial structure was characterized by historically changing mobilizations of discursive, symbolic, socioeconomic, and material processes that enrolled the physical qualities of H_2O and engineered the hydro-social constellation to make it act—albeit not always successfully—in a manner adequate to the dreams and aspirations of its master architects. The assembling of historically and geographically changing networks of interests shaped and plied the hydro-social landscape in ways that embodied and reflected the power geometries and choreographies of the time. The making of Spain's hydraulic geography was predicated upon folding and forging a historically variable scalar geometry that often combined and enmeshed local networks with national and transnational connectivities. These alliances effectively marginalized or repressed those who dissented while nurturing the heterogeneous interests of those who took an active role in sustaining the existing networks and power relations.

I excavated first the origins of Spain's early-twentieth-century modernization process (1890–1930) and the production of a national modernizing imaginary as expressed in debates and actions around the hydro-social condition. This was a period marked by a desperate attempt to pass through the trauma of the end of empire and to engage directly with Spain's troubled internal socio-ecological conditions. Great dreams and aspirations in a context of mounting political conflict and social struggle in the early decades of the century culminated in Spain's first authoritarian regime in the 1920s, which would lay the foundations for the subsequent decades of radical change. The prophylactic qualities of nature and nature's waters would offer the nexus around which narratives and projects promising redemption, modernization, and development became articulated. In chapters 5 and 6, I explored how Spain's modernization process during Franco's regime (1939–1975) became a specific scalar state project, framed around the production of a concrete techno-natural hydraulic edifice. Here, the focus shifts from documenting the failures to materialize the early twentieth-century hydromodernizing dream to considering how hydro-development became part of fascism's postwar vision for a fecund Spain, and how this "wet dream" was gradually put in place. Indeed, under Franco's totalitarian rule and with more than a little help from newly found transatlantic geopolitical friends, Spain's socio-ecological configuration was profoundly reengineered. In the final two chapters, I chronicled the joined process of democratic transition and transformation of Spain's hydro-structuralism into a new hydro-social assemblage that centered on the desalination of seawater for managing hydro-scarcities. The extraordinary process of rapid democratization after the demise of fascism and the exuberant cultural and social transformations that followed also inaugurated a series of new hydro-social imaginaries, discursive practices, and political-ecological conflicts. In this process, the "old" hydro-social edifice was increasingly contested and reassembled in new ways, extending the scalar choreography of hydro-social circulation into the sea in parallel with the recalibration of transnational and global configurations. Technocratic arguments, combined with a discursive and material reframing of the place and role of water, fused with pervasive processes of neoliberalization, intense regional conflict, and mounting environmental concern. Nonetheless, the ultimate objective remained the same, namely, determining how to make sure water keeps flowing so that both economic growth and the associated social and economic claims to water can be sustained.

Spain's hydro-social and techno-natural landscapes express simultaneously heroic modernizing desires, the legacy of a brutal authoritarian regime, the imprint of the elites' dreams, and the pain and suffering of millions of anonymous workers and peasants. It is from within this edifice, in the interstices of often enduring power assemblages, that new socio-ecological movements, innovative political visions, scalar arrangements, and alternative socio-technical projects were imagined, debated, framed, envisaged, and fought for. Along with changing geometries of power, water was rescripted and reimagined, reflecting transformations in ways of knowing, talking, sensing, seeing, and understanding what water is, why and how it matters, and how it acts.

Understanding Water

The relationship between water in its variegated acting and the nature of the associated social power relations have been and still are key concerns of many environmental historians, political scientists, engineers, policy experts and managers, cultural theorists, geographers, and sociologists, each foregrounding their particular take on what water is and what it does, and vying for the importance of their claims to "knowing" the truth about H_2O. The multiple narratives that pattern the stories in this book engage a range of disciplinary perspectives that have been too often enclosed in different and mutually exclusive disciplinary boundaries and epistemic communities. I chose not to foreground one particular perspective pulled from the shelf of available approaches and frameworks. Instead, I let the narrative do the hard work, as I believe that theory is forged through the slow work of painstaking research that permits the reconstruction of the futures of the past. This is not to say that no alternative narratives are possible or that there are no glaring and even impermissible gaps, silences, and absences in my work; I could have said more about groundwater, about the urbanization of water, about floods, about water and pleasure. I chose not to, hoping that others might walk down these paths. Rather, I intended to make theory come alive, to show how historical materialism, political ecology, science and technology studies, environmental justice arguments, and other materialist perspectives can be put to work and how, through the hard labor of storytelling, new theoretical insights can emerge.

Despite important differences, the preceding perspectives share a common theoretical concern that revolves around "why, where, and how does nature matter politically?" How does the stuff of nature enter the terrain of public concern and social agency and what does that, in turn, signify for earthly life, for rendering the socio-ecological predicament we are in more intelligible, for nurturing a socio-ecologically different, yet just and equitable, politics of the earth? The argument in this book also inscribes itself in this concern. The book can be read as a heuristic device that may provide ideas, insights, connectivities, narrative techniques, and methodological pointers for undertaking political ecological analysis in different historical and geographical contexts, and possibly with a meeting point other than water.

In an attempt to move away from dualistic interpretations of the relationship between nature and society, but still insisting on the performative role of nature's matter, the flow of the argument here maintained that, while there is a strong relationship between the changing nature of both state and society on the one hand and the modes of transforming and producing new techno-natural arrangements on the other, this relationship is mutually constituted and coevolves. It is historically and geographically variable and contingent; intrinsically bound up with often radical and contested imaginaries, cultural practices, engineering expertise, and shifting political-ecological power relations. The ultimate outcome of this process cannot be discerned in advance; it is the result of choreographies of contestation and paths of struggle. The materiality of the physical environment and its dynamics do not function primarily as an external given, but rather as historically constituted, materially and socially coproduced, environments. It is precisely in and through the contested production of new hydro-social environments that new forms of state organization and social relations are forged. Water and its flows are a synecdoche for society as a whole, one that captures its relative coherence as well as the forces that continuously undermine its stability and strive for change and transformation.

Politicizing Environments: Or . . . "Water Does Not Exist!"

While my gaze was firmly focused on Spain and its water, the intellectual focus of my argument was resolutely fixed on understanding water as a political category. In doing so, I departed from technocratic and managerial approaches to water issues and inserted water squarely into the terrain of political conflict and social struggle. Relational and territorial notions of socio-natural ordering were mobilized in the context of an analysis that aimed at considering how nature is remade through the fusion of the social, the technical, and the physical. The political and the technical, the social and the natural, became mobilized through and etched in spatial arrangements that shaped distinct and multiscalar geographies and landscapes—landscapes that celebrate the desires of elite networks, reveal the scars suffered by the disempowered, and nurture the possibilities and dreams for alternative visions.

The political ambition of the book was to show how socio-ecological configurations (in this case the hydro-social cycle) become constituted through a process of convening humans and nonhumans as they become enrolled in socio-ecological assemblages, and how the dynamics of disassembling and reassembling express shifting imaginaries, dreams, and political-economic and social power relations. The focus was on how convening is inevitably also a process of convoking, and on who or what does the summoning. Most important, I intended to show how nature's mobilization is always a question of contentious discursive enchainment, foregrounding particular materialities of nature, political contestation, and socio-environmental struggle. The variegated manners in which nature can be and is mobilized suggest ultimately that different socio-ecological constellations are always possible, up for grabs, for the making. What and how they will be made depends on who or what decides the political choreographies of their making. The diverse geographies of the world can and should indeed be understood as the outcome of an intricate socio-environmental process that perpetually transforms the socio-physical metabolism of nature. Mobilizing nature, although usually portrayed as a technological and engineering problem is, in fact, as much part of the politics of life as any other social process. The recognition of this political meaning of nature is essential if environmental processes are to be combined with just and empowering development.

As I showed throughout the book, nature and its waters do not exist outside the metonymic chains and social practices that offer some sort of instable meaning of what nature or water is and around which living environments become politicized. There are all manner of environments and assemblages of socio-natural relations possible and feasible. Nature's acting is variegated and heterogeneous, and its shifting discursive presentations or imaginary representations reflect these heterogeneities. Consider, for example, how water can and has been imagined as "good" or "bad", "just" or "unjust," "scarce" or "abundant," "source of life" or "causing disaster," "private" or "common." All socio-spatial processes are invariably predicated upon the circulation, metabolism, and summoning of particular social, cultural, physical, chemical, or biological processes, and their outcome is contingent, often unpredictable, and always immensely varied and risky. These processes produce a series of both enabling and disabling socio-environmental conditions (Heynen, Kaika, and Swyngedouw 2006). Processes of metabolic change are, therefore, never socially or ecologically neutral. The unequal ecologies associated with uneven property relations, the impoverished socio-ecological life under the overarching sign of commodity and money, and the perverse exclusions choreographed by the dynamics of uneven eco-geographical development at all scales suggest how the production of socio-ecological arrangements is always a deeply contentious and conflicting, and hence irrevocably political, process. Therefore, the production of socio-environmental arrangements implies fundamentally political questions, and has to be addressed in political terms.

It is in this sense that I claim, with Slavoj Žižek, that "Nature [and Water] does not exist!" (Žižek [1992] 2002; Swyngedouw 2010b). For Žižek, any attempt to stabilize and suture the meaning of nature, to inscribe a particular set of symbolizations in its name, is a decidedly political gesture. The disavowal or the refusal to recognize the political character of such gestures, and the attempts to universalize the situated and positioned meanings inscribed metonymically in Nature or Water lead to perverse forms of depoliticization, to rendering Nature politically mute and socially neutral (Swyngedouw 2007). It is precisely the recognition of the inherent slipperiness and multiplicities of meaning suggested by different metonymic enchainments of really existing things, emotions, and processes that urges us to consider that perhaps the very concept of nature itself should be abandoned. This book and many others on related themes indeed have shown conclusively that nature outside the social and the political does not exist (Morton 2007; Swyngedouw 2010b). What remains an enigma nonetheless is how we, both in our everyday life and mundane policies as well as in hydrological and engineering science communities, continue to see and act on nature and water as if we do not know this.

The fact that access to and distribution of water is highly uneven is well known. The articulation of the use and techno-natural transformation of water with social and political processes in which actors take highly unequal positions is also well documented. Nonetheless, both popular and scientific arguments remain predominantly fixed on water as a thing in itself and on how its variegated natural acting—even if the vital role of humans is acknowledged—constitute the determinant of our hydro-social condition. Droughts, water scarcity, hurricanes and floods, river flows and aquifer dynamics, thirsty lands and cities, technical infrastructures and distribution systems persist as the privileged entry through which the water conundrum is conventionally approached. Despite detailed scientific analysis of and sophisticated insights into the key social drivers and bottlenecks that structure the aquatic edifice, the simple fact remains that too many people still die prematurely or suffer unnecessarily because of water-related conditions that would be relatively easy to remedy had it not been for uneven power relations and perverse geographies of uneven development. Water keeps flowing uphill, to money and power.

Although we do really know that water injustices and inequalities choreograph the world's diverse hydro-social constellations and that struggles over water intensify, we rarely act on the basis of these insights. It seems indeed that ideology today functions precisely as the disavowal or foreclosure of what is already known. As Slavoj Žižek puts it: "We know very well how things are, but still, we act as if we do not know" (Žižek 1989). In a context of proliferating accumulation by dispossession, of unchecked concentration of resources in the hands of the few—often nurtured by managerial objectives that consider the techno-managerial organization of optimal market forces as the only horizon of the possible—and of rapidly deepening unequal social, political, and economic power relations, all manner of socio-ecological struggles that revolve around the signifier of "justice" are nonetheless actively resisting the often violent appropriation, not only of water, but also of a wide range of other common-pool resources too. These struggles, despite their radical heterogeneity, share a concern with a more equitable and solidarity-based organization of access to and appropriation and transformation of the commons. They signal how water is indeed thoroughly political and politicized.

I would like to suggest that opening up the debate over water-as-commons organized through an egalitarian and therefore democratic being-in-common (as the form of political organization of the social) might permit shifting the terrain somewhat from the currently dominant ethical concern with "justice" and an analytical focus on struggles of resistance to more directly and openly political visions and imaginaries that might nurture and galvanize politicized struggles aimed at a more egalitarian transformation and collective management of the commons of the earth. Such a scholarly perspective and situated political position would move away decidedly from considering water as a predominantly techno-managerial concern to one that focuses squarely on socio-biological life and well-being. This, of course, presumes a perspective that does not ignore or disavow radical contestation, explores mutually exclusive perspectives and imaginations, and acknowledges the profoundly varying social and political power positions of the interlocutors in the process. Achieving equitable socio-ecological governance implies a form of democratic politics that includes considering different political constellations of organizing the hydro-social cycle. It points inevitably toward a perspective that destabilizes consensus-based models that can presumably be assessed neutrally on the basis of scientific validity, efficiency, productivity, and inclusiveness. It is precisely such a mode of consensual techno-managerial management within an assumedly undisputed frame of market-led efficiency that reproduces the existing water inequalities. The water conundrum is indeed an emblematic issue, one that expresses in its variegated meanderings the functioning of political democracy, not just as a system of governing, but also as a set of principles articulated around equality, freedom, and solidarity.

Democratizing environments become, therefore, an issue of enhancing the democratic content of socio-environmental construction by means of identifying the strategies through which a more equitable distribution of social power and a more inclusive mode of producing socio-natures can be achieved. This requires reclaiming proper democracy and proper democratic public spaces (as spaces of agonistic dispute) as a foundation of and condition for more egalitarian socio-ecological arrangements. This also requires the naming of positively embodied egali[ber]tarian socio-ecological futures that are immediately realizable. In other words, producing egalitarian political-ecologies are about demanding the impossible and realizing the improbable.

Notes

1 "Not a Drop of Water..."

1. I define imaginaries as the combination of particular imaginations, dreams, fantasies, and images.

2 The Hydro-Social Cycle and the Making of Cyborg Worlds

1. Translation:

The Duero crosses the oak heart
of Iberia and of Castile.
                Oh, sad and noble land,
that of the high planes, barren and rocky,
of fields without ploughs, neither lakes nor groves;
crumbling cities, roads without inns,
...
Miserable Castilla, yesterday dominating,
wrapped in rags, disdaining the unknown.

3 "*Regeneracionismo*" and the Emergence of Hydraulic Modernization, 1898–1930

1. The promise of resurrection resides in nature.

2. *Caciquismo* refers to the general practice of systematic electoral manipulation, vote rigging, and vote buying at the local level by political bosses who are usually also the local economic elites (see Valera Ortega 2001; Ramírez Ruiz 2008).

3. This remark was already used in the sixteenth century, allegedly first made by Fray Francisco de Ugalde to Charles V, Holy Roman Emperor and king of Spain. Francis Bacon writes about Spain: "Both the East and the West Indies being met in

the crown of Spain, it is come to pass, that, as one saith in a brave kind of expression, the sun never sets in the Spanish dominions, but ever shines upon one part or other of them: which, to say truly, is a beam of glory" (Bacon and Montagu 1825, 438).

4. "Este país de obispos gordos, de generales tontos, de políticos usureros, enredadores y analfabetos, no quiere verse en esas yermas llanuras . . . donde viven vida animal doce millones de gusanos, que doblan el cuerpo, al surcar la tierra con aquel arado que importaron los árabes" (de Maeztu [1899] 1997, 109).

5. "*La invertebración*" is difficult to translate into English. For the intellectual and philosopher José Ortega y Gasset, *la invertebración*—the lack of integration as a geographical and moral-political project—is the main theme he introduced in his 1921 book *España Invertebrada* (Ortega y Gasset [1921] 2007).

6. "¿Y qué se encuentra en la inmensa meseta que se extiende desde Jaén hasta Victoria, desde León hasta Albacete, desde Salamanca hasta Castellón, desde Badajoz hasta Teruel? . . . ¿Qué es hoy Castilla? Recórrase en cualquier dirección. ¿Qué es hoy Castilla? Un páramo horrible poblado por gentes cuya cualidad característica aparente es el odio al agua y el árbol; ¡las dos fuentes de futura riqueza!"

7. "La pobreza de nuestro suelo—inmensa meseta central estéril, salpicada de algunos oasis y bordeada de una faja de tierra fértil—y la inclemencia de un cielo casi Africano."

8. "España era en buena parte un gran hinterland olvidado. En su interior escondía seguramente terribles cuadros de degeneración y pobreza que aguardaban ser redimidos, pero también inexplotados fuentes de riqueza, que la técnica moderna podía poner on producción, y maravillas legadas por la naturaleza y la historia, cuyo inventario permitiría devolver su integridad a los anales de la memoria nacional. Como en una verdadera empresa colonial, había un lugar reservado para el reconocimiento geográfico, científico y estadístico del territorio, con su doble promesa de prestigio cultural y de información con valor estratégico. También se trataba de facilitar la extensión de la presencia estatal, como agente de civilización y control, allí donde antes no llegaba su red capilar."

9. For a full bibliography, see Cheyne 1972.

10. "La desgracia de España ha nacido principalmente de que no llegó a entrar en la conciencia nacional la idea de que la guerra interior contra la sequía, contra las rugosidades del suelo, la rigidez de las costas, el rezago intelectual de la raza, el apartamiento al centro europeo, la falta de capital, tenía una importancia mayor que la guerra contra el separatismo cubano o filipino, y no haber sentido ante ellas las mismas alarmas que sintió ante ésta, y no haber hecho por la una los mismos sacrificios que no vaciló en hacer por la otra, de no haber confiado a los ingenieros y los maestros el raudal de oro que ha prodigado, triste suicida, a los almirantes y generales."

11. "No hay clima tan benigno como nuestro clima, ni cielo tan próvido como nuestro cielo, ni suelo tan fértil y abundante como el suelo de España; aquí, la Naturaleza provee generosamente al sustento del hombre casi sin esfuerzo . . . los demás pueblos se orarían de hambre si nosotros no les ofreciéramos las sobras de este festín espléndido á que nos tiene perpetuamente convidados la Naturaleza."

12. "Que nuestro clima es de los peores, nuestro suelo de los menos fértiles, nuestro cielo de los más ingratos y avaros, nuestra vida de las más penosas y difíciles, nuestro pueblo de los más hambreados y astrosos, nuestra lengua de las más pobres, nuestro ingenio de los menos fecundos, nuestra participación en la obra común del progreso humano de las más nulas; que no hay tierra en Europa que menos se parezca á una Jauja que la tierra española. . . . Si en otros países basta con que el hombre ayude á la Naturaleza, aquí tiene que hacer más: tiene que crearla."

13. "La planicie central, y acaso la mitad de España, es una de las regiones más secas del globo, después de los desiertos de África y Asia. Provincias hay, como Murcia . . . donde apenas si ve una nube en todo el año . . . los corrientes atmosféricas del Mediterráneo y del Atlántico no vierten sobre los abrasados campos de la Península toda el agua que necesitan las plantas para vegetar y fructificar; pero hay inmensos depósitos de ella en las crestas y en las entrañas de los montes, y podemos derramarla con la regularidad matemática de las pulsaciones sobre el país, cruzándola de un sistema arterial hidráulico que mitigue su calor y apague su sed."

14. Later, he expanded the theses developed in this paper in *Las Malas de la Patria* (The Evils of the Fatherland) (Mallada 1890).

15. The *Institución Libre de Enseñanza* was established in 1876, partly as a response to the dogmatic and censored official educational system, and would play a key intellectual role in Spain until the Civil War. Its mission was to educate without reference to any official dogma in religious, political, or moral affairs. Many regenerationists were prominent members of the Institute.

16. "Para rehabilitarnos, imprescindible es comenzar rehabilitando la propia tierra: condición esencial y absoluta. Y para restaurar ese medio geográfico, que es nuestra patria, quién negará que lo primero ha de ser conocerlo bien y exactamente? De aquí el carácter verdaderamente patriótico con que el cultivo de la geografía ibérica ha de ofrecerse a los ojos de todos los españoles."

17. "No hay nada más urgente para nuestra reconstitución nacional que un profundo estudio de nuestra geografía y nuestro suelo, que será el germen del gran renacimiento político de España."

18. The dominant view among the regenerationists was that nothing in Spain could change as long as the land-owning oligarchic aristocracy stayed in power via a system of *caciquismo* that assured a solid hold of traditional forces on the various levels of the state apparatus (see Costa [1901] 1998; Martínez Alier 1968; Pérez 1999).

19. "La Política Hidráulica entendida en un sentido amplio y simbólico, como un proceso de transformación acelerado de la agricultura de extensiva y tradicional en moderna e intensiva, debe constituir el vector fundamental de la política nacional, catalizando una reforma agraria que posibilite un desarrollo económico equilibrado y evite el progresivo proceso de proletarización de las masas campesinas, moderando la polarización social y la lucha de clases."

20. Dams or Death!

21. "'Política hidráulica' es una locución trópica, especie de sinécdoque que expresa en cifra toda la política económica que cumple seguir a la Nación para redimirse . . . una expresión sublimada . . . de la política económica de la Nación."

22. In the 1980s, this mission to create a national artery system, first formulated at the turn of the century, would become the backbone of the highly controversial Draft Proposal of the Second National Hydrological Plan, which proposed to link all of Spain's river basins together by means of a national, large-scale system of interregional water transfers (see chapter 7).

23. "Hay países que . . . solo, únicamente, exclusivamente, pueden ser países civilizados a costa de esa política hidráulica desarrollada en las magnas obras precisas al efecto. España entre ellos. . . . Y la verdad es que las agricultura civilizada Española . . . se halla férreamente sujeta a este dilema implacable: o tener agua o perecer . . . Se impone, pues, la política hidráulica, esto es, la conversión de todas las fuerzas nacionales hacia esa gigantesca empresa. . . . Hay que atreverse a restaurar magnos lagos, verdaderos mares interiores de agua dulce, multiplicar vastos pantanos, producir muchedumbre de embalses, alumbrar, aprovechar y detener cuantas gotas caen dentro de la península sin devolver al mar, si se puede, una gota sola."

24.
The Land of Soria is arid and cold.
Over the bald hills and mountains,
Green meadows, ashen hills
. . .
The earth does not revive, the field dreams
Snow-covered in early April.

25. Man and land make each other.

26. "El otoño habíase pasado sin una gota de agua, hasta el punto de haberse hecho la sementera en seco. Las lluvias de invierno faltaron asimismo. Ni un solo día habían dejado de reinar los cierzos del Nordeste, secos y helados. Y qué heladas! Negras, abrasadoras, sin una vesícula de vapor de agua en la atmósfera, con temperaturas de doce y quince grados bajo cero."

27. "Sin la previa solución del problema vital de los riegos, no hay en este país reformas posibles de importancia."

28. The quotes from Unamuno and de Maetzu are taken from Driever 1998b.

29. "Destrozando a los caciques, acabando con el poder de los ricos, sujetando a los burgueses. . . . Entregaría las tierras a los campesinos . . . y mi dictadura rompería la red de la propiedad, de la teocracia."

30. "Hoy Castro Duro ha abandonado ya definitivamente sus pretensiones de vivir. . . . Las fuentes se han secado, la escuela se cerró, los arbolillos . . . fueron arrancados. La gente emigra . . . , pero Castro Duro sigue viviendo con sus veneradas tradiciones y sus sacrosantos principios . . . dormido al sol, en medio de sus campos sin riego."

31. "La agricultura castellano lo que necesita es, no la política de los leguleyos y de los vividores, sino la política hidráulica . . . los riegos . . . el cultivo intensivo."

32. For an analysis of the stability of the *latifundistas* and the associated political order, see Costa [1901] 1998, Martínez Alier 1968, and Varela Ortega 1977.

33. "Y es que fuera de las cuencas de los ríos, todo está muerto, abrasado, desnudo, desmenuzado y polvoriento. Es que la tierra se muere de sed; es que no hay árboles; y como no hay árboles no hay agua ni vida, y en todas partes se encuentran horrores que parecen producidos por algún cataclismo geológico."

34. "Lo necesario no es política hidráulica, que no significa nada, sino política forestal que significa mucho."

35. "La iniciativa y la intervención del Estado para realizar numerosas obras de pantanos y canales de riego, es absolutamente indispensable si ha de darse á la riqueza agrícola é industrial del país el impulse regenerador que con tanta insistencia como empeño reclama la opinión en estos momentos en que, perdidos los territorios que un día se consideraron como base de la riqueza de la Metrópoli, hay que volver los ojos sobre el suelo patrio para sacar de él cuantos recursos ofrece, que han estado abandonados durante muchos siglos."

36. "Recurrir a las grandes alineaciones orográficas para efectuar una división de la superficie terrestre representa una aportación realizada desde el estricto campo de nuestra disciplina y muestra al mismo tiempo, por lo menos inicialmente, el abandono de las divisiones políticas y la importancia concedida a otros enfoques y conceptos."

4 Chronicle of a Death Foretold

1. I use this heading in a way that mimics Karl Marx's used of it. "De te fabula narratur" can be translated as "Of you the tale is told." By this Marx meant that the future of Germany was already visible across the English Channel. In this context, I mean that the future of Spain after 1939 until today (what I discuss from chapter 5 onward) was already visible/prefigured in what happened during the early decades of the century. Marx takes this quotation from Horatius.

It can and may also be interpreted in the sense that what happened in Spain can also be found and identified elsewhere. That is, for readers of the book who come from other countries, the story of torment told here about Spain finds parallels in many other countries.

2. Ley sobre Dominio y Aprovechamientos de Aguas (August 3, 1866).

3. Ley de Aguas (June 13, 1879).

4. Ley de Base de Obras Públicas (November 14, 1868; Ley de Canales de Riego y Pantanos (February 20, 1870) and Ley Gamazo (July 27, 1883).

5. A Royal Decree of 1890 established the Agricultural Chambers as a public-private partnership with the objective of organizing the farmers, facilitating the relationship with the state, and promoting the interests of farmers and the development of agriculture.

6. For further reading on signifying chains and the imaginary institution of state and society, see Castoriadis 1998; Kaika 2010; Swyngedouw 2010b.

7. "Hemos empezado la guerra de la paz, la guerra del trabajo, la lucha del progreso, que en lugar de devastar, restaura; de destruir, construye; de esquilmar, enriquece."

8. España ha varios siglos ambicionó más espacio, y á manera del líquido que para ganarlo necesita convertirse en vapor, consumiendo energía calórica, así cuanto más se extendió, más energía agotó de la metrópoli; luego, la frialdad de la ingratitud ha vuelto á condensar el vapor, y al concentrarse las moléculas otra vez en su recipiente primitivo devolverán aquellas energías llevadas, que hoy todo español empleará en la transformación de su patria.

Después de las hondas convulsiones sufridas, dirige nuestro país su angustiosa mirada hacia todo elemento de vida que pueda restaurar esas perdidas energías y hoy cifra sus esperanzas en el riego. Pero, aun flotando estas aspiraciones en el ambiente, se necesitaba encauzarlas, era preciso polarizar todos los ánimos, orientar todas las fuerzas y aplicarlas al mismo objeto en un momento dado para convertirse en potente impulsión nacional.

9. "Nuestras más renombradas vegas serían susceptibles de un cultivo más intensivo si pudiesen aprovechar en el estío una parte siquiera de las aguas que en las demás estaciones del año corren presurosas ante la vista del agricultor á perderse en la inmensidad de los mares, cuando no le causan destrozos y aun grandes desastres. . . . Al Estado corresponde, pues, tomar á su cargo el estudio y la construcción de un plan de pantanos y canales de riego; y el Gobierno que trata de abrir una era de regeneración en esta desdichada tierra . . ."

10. "Para La Nueva Política. Los Canales de Riego," *El Imparcial*, Madrid, April 6, 7, 8, 1899.

11. "La obra de embalsar y detener las aguas de nuestros ríos, que con velocidad torrencial y devastadora salvan la distancia que media entre los manantiales y el

mar, es gigantesca, y como tal, está rodeado de abrumadores obstáculos. Existen, si, aunque no insuperables para una laboriosa perseverancia . . . pero si sólo hubiesen de acometerse los empeños de poca monta, los asuntos llanos y sencillos, nunca remediaríamos defectos subsanables de nuestro suelo. Ni jamás elevaríamos nuestra desmedrada y enteca producción."

12. DD.SS. Congreso, nr. 33, July 11, 1899, 821–824. The declaration was presented by Rafael Gasset, Vicente Alonso Marínez, Gumersindo de Azcárate, Francisco de Federico, José Canalejas, Marqués de Figuerora, Ezequiel Ordóñez.

13. DD.SS., 823–824.

14. "No ceden, de cierto, en elocuencia los datos geográficos á los que aporta la Historia. El Sahara nos envía por la costa levantina vientos asoladores, soplos de fiebre del continente africano, que todo lo secan y esterilizan. Las aguas pluviales aparecen, ya que no escasas, distribuidas con tal injusticia é irregularidad."

15. "Una acción energética que acabe con esta política de palabras, que nos conserva hoy tan pobres y tan débiles como el día que se firmó el Tratado de París."

16. "Ahí están varios ingenieros, y no pequeña parte del personal subalterno, cruzados de brazos por disposición del Gobierno, y ahí están también diversos ríos que, tras indolente carrera, se lanzan á la mar, ó se ocultan vergonzosamente en sus propias arenas."

17. "Precisamente cuando el país entero cree que en ello pudiera encontrarse uno de los medios más adecuados para promover de una manera práctica y eficaz la riqueza nacional, contrarrestando las dolorosas pérdidas que España ha sufrido últimamente."

18. "Alfa y Omega de la faena que se ha impuesto el Directorio Militar es acabar con la vieja política. El propósito es tan excelente que no cabe oponerlo reparos. Hay que acabar con la vieja política."

19. See Caprotti 2007 and Caprotti and Kaika 2008.

20. After the fall of the dictatorship, he went into exile in Argentina, returned to Franco's Spain (in whose government his brother was minister and governor of the Bank of Spain), and took up a variety of high administrative offices (among others, president of RENFE, the national train company).

21. "Todos los servicios de carácter hidráulico deberían estar centralizados . . . en una misma dependencia para cada Cuenca general, único medio de que su eficacia respondiera al interés nacional de esa importantísima fuente de riqueza."

22. "La vigorosa semilla de la Confederación traía escondido entre sus pliegues el germen, el virus letal de su propia destrucción."

23. "No ha sido nunca ni podía ser la Confederación patrimonio exclusivo de un partido, ni siquiera de un régimen político. Fuese cualquiera el momento en que

recibió su existencia legal, es una Institución que nació y vivió para atender las necesidades nacionales y para fomentar e intensificar la riqueza pública, fin común a todas las doctrinas y a todos los procedimientos de gobierno."

24. The courts further investigated the accusations of malfeasance by the Ebro Confederation. During the Popular Front government, Pardo was relentlessly prosecuted and his house in Madrid ransacked by republican forces. He was accused of supporting the Catalan nationalists who demanded full control over "their" waters. In August 1936, under threat of prosecution, he sought refuge in the Chilean embassy and fled one and half years later to Paris with his wife and three children. He would later return to Spain, was rehabilitated, and took up the position of Inspector of Public Works and President of the National Council of Public Works in the Franco administration (Marcuello 1990). During his period of exile in Paris, Pardo wrote a major defense of the great achievements he and his colleagues accomplished in the Ebro Valley, while defending the need for local autonomy of the river basin authorities against partisan interest in order to implement the great works required for the nation as a whole (Lorenzo Pardo 1931).

25. For the full text of the plan, see http://hercules.cedex.es/informes/planificacion/1933-plan_nacional_de_obras_hidraulicas/default.htm (accessed December 20, 2010).

26. "Sus enormes posibilidades productivas en espera de un caudal nuevo que las actualice y las incorpore a la economía nacional, planteando así la necesidad de recurrir a recursos externos para abordar y resolver definitivamente el problema del sureste peninsular."

27. "El fantasma del trasvase . . . se desvanece ante el soplo poderoso de una suprema conveniencia nacional. Si un día somos dueños de las aguas españolas, que caen en cantidad suficiente, aunque desigualmente repartidas, y las podemos distribuir según esa conveniencia, habremos realizado el acto más firma de soberanía. La fina y melancólica espiritualidad del Norte, templada en las austeridades castellanas, podrá llegar a fructificar aquí en Levante por la conjunción de aquellas aguas y este sol, y entonces sí que serían los vuestros unos productos nacionales capaces de llevar al último rincón del mundo, desde América hasta el Extremo Oriente, una encarnación del alma española."

28. "Sus trabajos fundamentales y básicos serán de gran utilidad, y creemos que deben continuar y ponerse al día, orientándolos convenientemente."

5 Paco El Rana's Wet Dream for Spain

1. "A la generación llamada del 98—pensadores y 'diletantes'—se ha opuesta la generación de los hombres de acción surgidos desde 1935, cuyas realizaciones se ha traducido en el desarrollo económico de España."

2. "En los decenios posteriores al Plan de 1940 realmente se *crean* los ríos españoles, construyéndose o iniciándose las principales obras de regulación, sobre las que se apoya cualquier política nacional integrada de los recursos de agua."

3. "Estamos Dispuestos a que no se Pierda una Sola Gota de Agua ni Persista una Sola Injusticia."

4. "FRANCISCO FRANCO, CAUDILLO DE ESPAÑA, ORDENÓ SU CONSTRUCCIÓN. CON ÉL DOMINÓ LAS TURBULENTAS AGUAS DEL RÍO SEGURA PARA QUE FECUNDEN PACIENTEMENTE SUS SEDIENTAS TIERRAS Y REDIMIÓ A LOS HOMBRES QUE LAS TRABAJAN DEL TEMOR MILENARIO A LAS INUNDACIONES Y LAS SEQUIAS."

5. "Es el ritmo del corazón que da impulso a los corazones de todos los españoles redimidos de la sed, del hambre y del trabajo estéril en tierra dura gracias a la cordial política de Franco en quien todo esfuerzo cabe si es para despertar a España de su vieja modorra."

6. "Nos dolía España por su sequedad, por su miseria, por la necesidades de nuestros pueblos y de nuestros aldeas, y todo ese dolor España se redime con estas grandes obras hidráulica nacionales, con este pantano del Ebro y con los demás que en todas las cuencas de nuestros ríos van creándose, embelleciendo su paisaje y creando ese oro líquido que es la base de nuestra independencia."

7. "Esta eficaz política de regulación, que ha conseguido corregir en gran medida la desigual distribución en el tiempo de nuestra precipitaciones, logrando restablecer una regularidad en los regímenes naturales de los ríos para crear corrientes permanentes donde estrictamente tan sólo había torrentes, ha sido la base fundamental de la moderna planificación integral hidráulica española."

8. Traditionalist Spanish Front of the Juntas of the National-Syndicalist Offensive.

9. *El Cid* (Rodrigo Díaz de Vivar [1043–1099]), nobleman and military commander, is the mythologized national hero of Spain's struggle against the Moors and for Spanish unification.

10. BOE, March 10, 1938, Jefatura del Estado. Fuero del Trabajo.

11. "Renovando la Tradición Católica de justicia social y alto sentido humano que informó nuestra legislación del Imperio, el Estado Nacional en cuanto es instrumento totalitario al servicio de la integridad patria, y Sindicalista en cuanto representa una reacción contra el capitalismo liberal y el materialismo marxista, emprenda la tarea de realizer—con aire militar, constructivo y gravemente religioso—la Revolución que España tiene pendiente y que ha de devolver a los españoles, de una vez para siempre, la Patria, el Pan y la Justicia."

12. For detailed analysis of these alliances and consolidating networks, see among others, the magisterial books by Gallo (1974), Fusi and Palafox (1989), Preston (1990, 1995), Preston and Lannon (1990), Carr (1995, 2001), and de Riquer (2010).

13. Most of the literature on hydraulic policies during Franco focuses on irrigation and internal colonization. Expanding those was indeed the refrain endlessly repeated by state officials and Franco himself. For reviews, see, among others, Ortega 1975; Gil Olcina and Morales Gil 1992; Melgarejo Moreno 1995, 2000; Barciela López and López Ortiz 2000; and Rodríguez Ferrero 2001.

14. For a detailed analysis of Barcelona's and Madrid's water supply, see March Corbella 2010.

15. "No son los aguas que corren por las cordilleras y vertientes directas al mar, por bellas que sean, las que más rinden, sino las que se dejan dominar, las que se suman a los cauces comunes; ésas son las que dan toda la riqueza para la industria y para la agricultura. Y esto se puede aplicar no solamente a la cuestión hidroeléctrica, sino a todas las manifestaciones de la industria y del progreso en España. . . . La riqueza hidráulica que se pierde es de 21000 millones de metros cúbicos en el Mediterráneo y de 26000 millones hacia el Atlántico, y hay zonas pobrísimas y zonas con abundancia, y hace falta estudiar ya el traspaso de agua de unas zonas a otras. Es decir, que nuestro sistema hidrográfico no nos permite dividir por zonas y nos obliga mirar el conjunto, pues de otro modo no hay equilibrio en el país."

16. "Hoy hay que poner todas nos fuerzas, sin tibieza ni cobardía, al servicio de España y la economía española, en trance de destrucción por la traición fascista. Para levantaría, necesitamos a todos los hombres de buena voluntad. . . . Hoy se trata de vencer el enemigo. Mañana sonara la hora de construir una España modelo de países civilizados."

17. "La Revista de Obras Públicas reanuda hoy su publicación, interrumpida durante muy cerca de cuatro años. Al iniciarse el Glorioso Movimiento Nacional, todas las simpatías del Comité de redacción estaban del lado del Gobierno que venía a restaurar le eternal tradición española. Los rojos lo sabían y desconfiaban; esperaban quizá una adhesión a su causa, entusiasta y explícita, quizá arrancado por el terror, y como esta no llegaba tal como la deseaban, a los pocos días de ser publicado el numero de agosto de 1936, un grupo de indeseables irrumpió e nuestra Redacción, pistola al cinto, y se incautaron de la Revista a nombre del Sindicato Nacional de Arquitectura e Ingeniera de la Unión General de Trabajadores. Aun intentaron que el Comité de Redacción continuara sus trabajos bajo aquella improvisada dirección, pero ninguno de sus miembros se prestó a colaborar en tales condiciones, y pocos días después bajo un nuevo Comité, siguieron publicando algunos números, utilizando el original, en su mayor parte, por nosotros acumulado; pero cuando éste se agotó y ello ocurrió a los tres meses, cesaron en su empeño."

18. "De recuperar la vocación imperial de España para lo que propone acrecentar nuestra fe religiosa, haciéndola base de nuestras actuaciones y venero vivo de todas las virtudes raciales. Frente a las dogmas falsos de los podridas democracias, los lemas de nuestra Edad de Oro, ya incorporados por el Generalísimo. Frente a la libertad, el servicio. Frente a la igualdad, la jerarquía. Como superación de la fraternidad, la hermandad que presupone le común paternidad de Dios."

19. Lino Camprubí (2012) explored the relationship between engineers and the totalitarian transformation of river basins for the case of the northern Noguera Ribargorzana river basin.

20. "Un 'sentido Hidráulico,' que es como una especie de sexto sentido que permite intuir el comportamiento del agua en sus movimientos. . . . Para ser un buen ingeniero hidráulico hace falta saber, ver, oír y tocar el agua con los ojos, los oídos y las manos . . . con el alma vibrante. . . . No se sorprendente que en España hayamos tenido y tengamos magníficos ingenieros hidráulicos . . . porque España es el país de la imaginación por excelencia. . . . Si la historia abona la imaginación racial de los españoles, la Geografía y el clima no lo hacen en menor grado. . . . El ardiente sol que nos enjoya y reseca por cuanto contribuye a la escasez e irregularidad del agua en la Península. . . . Estas dos características conjugadas producían tal déficit de agua utilizable, que la lucha empeñada por los españoles para domesticarla, situándole de modo extenso y continuo a todo lo ancho de la Nación. . . . La frase evangélica de "por sus obras los conoceréis" . . . tiene adecuado aplicación a nuestros ingenieros hidráulicos. . . . A todos ellos, a los que pasaron y a los que hoy prosiguen su labor con tanta brillantez, la gratitud de España y la gratitud especial del Cuerpo de Ingenieros de Caminos."

21. See, for example, Mendiluce 1996.

22. "La caída de la monarquía precipitó aún más la catástrofe de nuestros centros de cultura, y la República lanzó a la Universidad por la pendiente del aniquilamiento y desespañolización, hasta el punto de que brotaron de su propia entraña las más monstruosas negaciones nacionales. . . La Ley, en todos sus preceptos y artículos, exige el fiel servicio de la Universidad a los ideales de la Falange, inspiradores del Estado, y vibra al compás del imperativo y del estilo de las generaciones heroicas que supieron morir por una Patria mejor. . . Al acometer esta empresa de transformación cultural y educativa se realiza la más fecunda e imperiosa consigna de la Revolución Nacional exigida por la sangre de los que supieron morir en acto de servicio y por la noble pasión de los que quieren ahora servir también con su vida a los supremos destinos de España."

23. See also Franco 1940, xvi–xvii.

24. See Boletin Oficial de l'Estado, Número 356, December 22, 1942. The obligation to screen them was lifted after the death of Franco in 1975.

25. All NO-DO reels have been made available digitally and can be accessed freely on the Internet. http://www.rtve.es/filmoteca/no-do/ (accessed June 21, 2014).

26. "El Caudillo de España, que en la horas de la guerra supo llevar nuestras tropas a la Victoria, es también el alma de esta labor reconstructora, con la que España cicatriza sus heridas, salvando todas las dificultades que openen las actuales circunstancias del mundo."

27. "Hemos venido a visitar vuestra provincial, a inaugurar varias obras transcendentes, a dar por terminadas una serie de ellas, y con ello a satisfacer la sed de vuestros campos, a regular vuestros riegos, que aumentarán el bienestar y multiplicarán la producción. . . . Necesitamos regar nuestros campos, dar satisfacción a la sed de tantas tierras. . . . Toda esa España es a la que hay que redimir sellando la hermandad entre las tierras y los hombres de España."

28. "Y así, con esta política de realidades, que puede sonar a tópico en las mentes deformadas, va cerrándose el cerco contra la miseria y la sed de los hombres y las tierras españoles, porque a la viaja esterilidad de los ríos el programa de Franco ha ido poniendo murallas de hormigón para hacer la luz, y venas de cemento para canalizar el agua y levaría a cumplir su fin de riego reparador de la vieja sed campesina. Y los campesinos, arracimados de júbilo, ofrecieron hoy a Franco la expresión de su gratitud sin límites, de su adhesión sin concesiones."

29. "Es el ritmo del corazón que da impulso a los corazones de todos los españoles redimidos de la sed, del hambre y del trabajo estéril en tierra dura gracias a la cordial política de Franco en quien todo esfuerzo cabe si es para despertar a España de su vieja modorra."

6 Welcome Mr. Marshall!

1. Although Villar del Río is a really existing village in the Province of Soria, the village scenes of the movie were filmed in Guadalix de la Sierra, located in the Province of Madrid.

2. Qué delito cometieron?

Sólo querían la igualdad
de los hombres y los pueblos.
A punta de pico y pala
hicieron ese Canal
calladitos y en silencio,
detrás estaba el guardián.
Canal del Bajo Guadalquivir:
que le quiten ese nombre,
que lo quiten por favor.
Es el Canal de los Presos,
lo hicieron con su sudor
. . .
Esto no es una poesía
es una ofrenda de honor
par todos de los que estuvieron
en campos de concentración.

3. Letter from President Roosevelt to U.S. Ambassador Norman Armour on American relations with Spain, March 10, 1945, released to press September 26, 1945, United States Department of State *Bulletin*; http://www.ibiblio.org/pha/policy/post-war/1945-09-26a.html (accessed April 5, 2011).

4. Of course, the position of the U.S. administration was not unanimous. Not surprisingly, the State Department shared the views of the United States' European allies, while the Pentagon and sections of Congress (the "Spanish Lobby," consisting mainly of Catholics, anti-communists, parts of the army, republican opposition to Truman, and business elites with interests in Spain) lobbied for a "normalization" of relations with Spain (Edwards 1999; Puig 2003). Public opinion was also very much divided. While the United States was officially nonintervening in the Civil War, more than three thousand U.S. citizens fought on the side of the Spanish Republicans during the war.

5. Like today, the official U.S. interpretation of these agreements very much follows the doctrine of the "manifest destiny" of the United States to redeem humanity from tyranny and dictatorships in whatever form they take historically. The 1953 pact is very much seen as such an initiative that would eventually pave the road for development, freedom, and democracy in Spain! In 1993, for example, in an article dedicated to the fortieth anniversary of the Madrid agreement, Vernon Walters, who acted as interpreter between Franco and both Eisenhower and Nixon stated: "The agreements signed in September 1953 . . . permitted the country [Spain] to become one of the leading economies of the world in an environment of freedom and democracy. . . . I believe that without the agreements between Spain and the United States, history could have been much different" (Walters 1993, 167).

6. "Si no hubiese sido por el esfuerzo realizado hace veintitrés años, España seria hoy comunista . . . en ese aspecto nos hallamos en deuda con el pueblo español: una deuda de gratitud. Hoy—añadió—todos sabemos que España es el país más amigo y aliado de Estados Unidos. . . . No hay motivo, pues, para negar a España su ganada derecho a participar plenamente en los sistemas defensivos occidentales."

7. "La Naturaleza concentró en ellas las posibilidades de aquellas grandes obras hidráulicas y de riego que hoy están cambiando la geografía española. . . . Ellos han sido la base firme de esta reconstrucción española, de este engrandecimiento de la Patria."

8. "Si las ideas Costanas se habían basado en la unidad de la cuenca como marco de realización de las propuestas hidráulicas, la planificación hidrológica amplia esta marco a escala estatal, al plantearse como uno de sus objetivos la posible corrección de los desequilibrios existentes en la península Ibérica, mediante la interconexión de las cuencas."

9. "El consejo de Ministros del 1 Junio 1955 se había examinando el más ambiciosa proyecto de la Historia de España: cambiar su estructura hidráulica. No se puede

seguir confiando a la lluvia estacional e irregular la prosperidad agrícola y industrial. Los pantanos construidos demuestran la eficacia del agua embalsada. Ahora se trataba de utilizar el Tajo como gran reserve para su redistribución, embalsando varios veces el mismo agua, distribuyendo y volviendo a distribuir en regadíos cada vez más lejano. . . . El 14 de Julio (1955), Franco inauguró las grandes embalses de Entrepeñas y Buendía, en la provincial de Guadalajara. Se les llamaba, orgullosamente y significamente, "mar de Castilla." Pues la próxima etapa en la política hidráulica—para realizar en decenas y con paciencia—consistía precisamente en el establecimiento de mares interiores con redes de transvase de agua."

10. "El Sureste libra una batalla denodada con el agua que se prolonga a lo largo de los siglos; esta historia, que está aún por escribir, tiene mucho de épica. La operación hidráulica que vamos a emprender es una gran empresa de justicia nacional."

11. With thanks to Joan Martínez-Alier for pointing me to this exemplary case.

12. The original film-reel can be found at http://www.rtve.es/filmoteca/no-do/not-837/1487132/ (accessed July 12, 2013).

7 Marching Forward to the Past

1. Hail, Hail, Hail, the cables have been cut!

2. See http://news.bbc.co.uk/2/hi/uk_news/484682.stm (accessed December 20, 2011).

3. ETA stands for Euskadi Ta Askatasuna (Basque Homeland and Freedom), the militant Basque nationalist and separatist organization.

4. "Voy a construir más Pantanos que Franco."

5. "La atención a los problemas cuantitativos del agua . . . debe ser permanente. No es posible de descansar en la tarea de acrecentar las disponibilidades hidráulicas. La experiencia española es sobradamente conocida, ya que después de un periodo de intensa actividad (1950–1970), parecía merecida una pequeña pausa (1970–1980); pero esta actitud ha traído como consecuencia situaciones de escasez que requieren hoy reforzar la actividad a índices superiores a los alcanzados en los períodos de mayor apogeo."

6. "Vertebración" refers to process of producing consistency, coherence, organization, and structure by means of a central spinal articulating configuration.

7. "El agua debiera ser una de nuestras preocupaciones fundamentales, y la política hidráulica habría de tener lugar preeminente entre las políticas territoriales."

8. "Además de todos los cambios económicos, sociales y jurídicos nombrados, continúan sin embargo existiendo los mismos problemas hídricos que fundaron en su momento, y que siguen fundamentando hoy, la existencia de una acción decidida

del Estado en el dominio público hidráulico bajo la forma de Planificación Hidrológico. . . . La presente ley contiene entonces una previsión de transferencias de recursos hidráulicos entre cuencas construida y orientado por un objetivo de gran ambición: se trata de sentar las bases, de una vez por todas, que permitan resolver la manifiestamente desequilibrada distribución de los recursos hidráulicos españoles."

9. The Ésera plays an important role in Costa's emblematic book *Política Hidráulica* (Costa Martínez 1911).

10. "El Ésera y muchos otros Éseras recorrerán la piel de España y sus aguas limpias serán, recordando el estilo poético de Costa, su sangre, su rocío y su oro, el camino de la liberación y de la opulencia colectiva. Por ello se formula el Plan Hidrológico nacional."

11. The General Agreement on Tariffs and Trade had become more forceful in imposing neoliberal reform on member states.

12. "Pero si, aun así, se plantea, inexorablemente, la necesidad de conectar cuencas, ha de tenerse en cuenta que las cuencas no son más que una división administrativa, ya que el agua es de todos los españoles, y así se recoge en la Ley de Aguas."

13. "Esta modalidad permite abrir nuevas vías a la participación de la iniciativa privada en el desarrollo de las infraestructuras hidráulicas, lo que posibilita una mayor responsabilidad de los particulares en la financiación de las infraestructuras y aprovechamiento y explotación de los recursos hídricos."

14. "Algunas zonas viven en la abundancia mientras otras tienen graves carencias. . . . En algunas partes se vive la confrontación social por un recurso natural: el agua; siendo ésta un bien común no se puede despilfarrar ni olvidar el deber solidario de compartir su uso. Las riquezas no pueden ser monopolio de quienes disponen de ellas."

8 Mobilizing the Seas

1. Programa A.G.U.A. stands for Programa Actuaciones para la Gestión y la Utilización del Agua, the legal framework for the management and use of water.

2. "The sea, an inexhaustible source of life." AcuaMed, http://www.acuamed.es (accessed July 27, 2011).

3. "Sin embargo, no podemos ver con pesimismo el futuro y mucho menos desde España, tan abundante en costas, porque, al fin y al cabo, el gran sucedáneo será el mar. El mar será nuestro gran recurso de reserva y, a pesar de ello, lo tenemos abandonado; su explotación es aún rudimentaria; para los hombres, para nosotros, es el gran desconocido. . . . Si los medios y los enormes esfuerzos humanos y económicos dedicados a la exploración de los astros . . . se hubieran empleado en conocer el mar y en obtener sus beneficios, creemos que los resultados hubieran sido harto más

fecundos ... Del mar podemos esperar grandes auxilios de todo orden, que apenas si se vislumbran."

4. "España desala el oro de sus costas."

5. "Quiero anunciar una nueva política del agua, una política que tomará en consideración tanto el valor económico como el valor social y el valor ambiental del agua, con el objetivo de garantizar su disponibilidad y su calidad, optimizando su uso y restaurando los ecosistemas asociados."

6. "Nuestro aliado más valioso para conseguir agua, donde hay escasez, es la desalación, la alternativa más económica, flexible y limpia para garantizar el suministro independientemente de las condiciones meteorológicas." http://www.acuamed.es/quienes-somos/que-es-acuamed (accessed October 22, 2014).

7. See http://www.water.siemens.com/en/municipal/Pages/water-scarcity.aspx (accessed August 23, 2011).

8. See http://www.acciona.com/news/acciona-agua-inaugurates-a-desal-plant-that-will-deliver-water-to-a-million-londoners- (accessed August 22, 2011); emphasis in original.

9. See http://www.zaragoza.es/contenidos/medioambiente/cajaAzul/19S5-P1-Jorge_Salas-PPTACC.pdf (accessed August 23, 2011).

10. See http://www.finance.veolia.com/press-release-710.htm (accessed August 23, 2011).

11. "La desalación es una fuente alternativa e inagotable de agua en un momento en el que el abastecimiento de este recurso se convierte en un verdadero problema en muchos puntos del mundo, agravado por la situación de cambio climático." http://www.abeima.es/corp/web/es/construccion/agua/desalacion/index.html (accessed August 23, 2011).

12. For detailed information on the Foundation for a New Water Culture, see http://www.fnca.eu/.

13. See http://cenews.com/post/722/headlines_from_around_the_web_ (accessed October 22, 2014).

14. See http://www.magrama.gob.es/es/ministerio/funciones-estructura/organizacion-organismos/otros-organismos-organizaciones/sociedades-estatales-del-agua/ (accessed October 22, 2014).

15. "A diferencia de otras comunidades autónomas, la Región de Murcia no cuenta con ningún partido nacionalista, ni se había dotado hasta ahora una imaginaria identidad nacional, ni se había afirmado frente a otras comunidades como un nosotros unitario y excluyente. Sin embargo, desde 1995, ha sido el PP, el partido supuestamente garante de la unidad de España, el que ha conseguido crear de la

nada un sentimiento patriótico de "murcianía."... Pero este nuevo sentimiento se asienta sobre una seña de identidad muy peculiar: el agua, convertido en el gran tótem comunitario de los murcianos mediante la simple y reiterada consigna del agua para todos."

16. "Por suerte, la Región de Murcia tiene una franja litoral que permite disponer de recurso primario infinito. Esta situación garantiza la eliminación de todas las incertidumbres o de disponibilidad que necesitan los trasvases, puesto que aplicando la tecnología se puede disponer de volúmenes parecidos a los que se estimaban transferibles del Ebro. . . . Ahora bien, la oferta de recursos provenientes de la desalación no puede aislarse de las otras aguas de la Cuenca, en el sentido de que recursos que antes venían a las zonas de costa desde aguas arriba, y que ahora deben de sobrar por proceder de la desalación, se pueden y se deben dedicar a abastecer zonas altas e intermedias de la cuenca. En otras palabras, la desalación debe contemplarse conjuntamente con todas las aguas disponibles del sistema."

References

Abad Muñoz, G. 2006. "Carta Abierta a la Ministra de Medio Ambiente, Dña. Cristina Narbona Ruiz." *Revista de Obras Públicas* 153 (3462): 47–48.

Acosta Bono, G., J. L. Gutiérrez Molina, L. Martínez Macías, and A. del Río Sánchez. 2004. *El Canal de los Presos (1940–1962). Trabajos Forzados: de la Represión Política a la Explotación Económica*. Barcelona: Crítica.

Aguilera Klink, F. 2008. *La Nueva Cultura del Agua*. Madrid: Los Libros de la Catarata.

Aguiló Alonso, M. 2003. "Ingeniería versus Obras Públicas en el Periodo de Entreguerras." *Revista de Obras Públicas*, no. 3434: 73–82.

Albaredo Page, E. 1871. "Divisiones Hidrológicas." *Revista de Obras Públicas* 19: 156–169.

Alcarez Calvo, A. J. 1994. "La Participación de los Usuarios en la Administración Pública del Agua." *Revista de Obras Publicas* 141 (3335) (September): 7–24.

Altamira, R. 1923. *Ideario Pedagógico*. Madrid: Editorial Reus.

Alvarez, A., and A. Del Campo. 1979. "Nuevas Presas a Realizar en España para Completar el Aprovechamiento Integral de los Recursos Hidráulicos." *Revista de Obras Públicas* 126 (3176): 1013–1020.

Angoustures, A. 1995. *Historia de España en el Siglo XX*. Barcelona: Editorial Ariel.

Antolín, F. 1999. "Iniciativo Privada y Política Pública en el Desarrollo de la Industria Eléctrica en España. La Hegemonía de la Gestión Privada, 1875–1950." *Revista de Historia Económica* 17 (2): 411–445.

Antolín Fargas, F. 1997. "Dotaciones y Gestión de los Recursos Energéticos en el Desarollo Económico de España." *Papeles de Economía Española*, no. 73: 193–207.

Areilza, J. M. de. 1984. *Memorias Exteriores, 1947–1964*. Barcelona: Planeta.

Armiero, M. 2011. *Rugged Nation. Mountains and the Making of Modern Italy*. Isle of Harris, UK: The White Horse Press.

Armiero, M., and W. Graf von Hardenberg. 2013. "Green Rhetoric in Blackshirts: Italian Fascism and the Environment." *Environment and History* 19 (3): 283–311.

Arrojo, F., and L. Visa. 2009. "Nueva 'Batalla' por el Agua del Ebro: La Generalitat propone abastecer del río a cuatro localidades que se hallan fuera de su cuenca, lo que provoca reacciones diversas en Aragón y Valencia." El País, December 1. http://elpais.com/diario/2009/12/01/catalunya/1259633239_850215.html (accessed October 20, 2014)

Arrojo Agudo, P., ed. 2001a. *El Plan Hidrológico Nacional a Debate*. Bilbao: Bakaez/Fundación Nueva Cultura del Agua.

Arrojo Agudo, P. 2001b. "Hacia una Nueva Cultura del Agua Coherente con el Desarrollo Sostenible." In *Ecología: Perspectivas y Políticas de Futuro*, ed. J. Araújo, 117–163. Sevilla: Junta de Andalucía-Fundación Alternativas.

Arrojo Agudo, P., ed. 2004. *El Agua en España. Propuestas de Futuro*. Guadarrama: Ediciones del Oriente y del Mediterránea.

Arrojo Agudo, P. 2005. *El Reto Ético de la Nueva Cultura del Agua: Funciones, Valores y Derechos en Juego*. Barcelona: Ed. Paidós.

Arrojo Agudo, P., and J. M. Naredo. 1997. *La Gestión del Agua en España y California*. Bilbao: Bakeaz.

Arroyo, A., and R. Boelens, eds. 2013. *Aguas Robadas. Despojo Hídrico y Movilización Social. Justicia Hídrica*. Lima: IEP; Quito: Abyayala.

Asían Peña, J. L. 1941. *Elementos de Geografía General e Historia de España*. 2nd ed. Barcelona: Bosch Casa Editorial.

Ayala-Carcedo, F. 1999. "De la Política Hidráulica a la Política del Agua Sostenible." *Tecnoambiente*, no. 90: 5–9.

Ayala-Carcedo, F. J., and S. L. Driever, eds. 1998. *Lucas Mallada: La Futura Revolución Española y otros Escritos Regeneracionistas*. Madrid: Editorial Biblioteca Nueva.

Azorín. 1982. *Obras Selectas*. Madrid: Editorial Biblioteca Nueva.

Bacon, F., and B. Montagu. 1825. *The Works of Francis Bacon. . . . A new edition: by Basil Montagu, Esq. [With a portrait from a miniature by Nicholas Hilliard.] L.P.* London: William Pickering.

Bakker, K. 2002. "From State to Market? Water *Mercantilización* in Spain." *Environment and Planning A* 34: 767–790.

Bakker, K. 2010. *Privatizing Water: Governance Failure and the World's Urban Water Crisis*. Ithaca: Cornell University Press.

Bakker, K. 2012. "Water: Political, Biopolitical, Material." *Social Studies of Science* 42: 616–623.

Baklanoff, E. N. 1976. "The Economic Transformation of Spain: Systemic Change and Accelerated Growth, 1959–73." *World Development* 4 (9): 749–759.

Baltanás García, A. 1993. "El Plan Hidrológico Nacional: Presentación." *Revista de Obras Públicas* 140 (3321): 7–18.

Bárcena Hinojal, I. 1999. "Itoiz: Actores y Teatro Comunicativo." *Soziologia Eta Zientzia Politikoaren Euskal Aldizkaria—Revista Vasca de Sociología y Ciencia Política*, no. 24: 143–154.

Bárcena Hinojal, I., and P. Ibarra Güell. 2001. "Itoiz: ¿Un conflicto local, nacional o global?" *Inguruak: Soziologia Eta Zientzia Politikoaren Euskal Aldizkaria—Revista Vasca de Sociología y Ciencia Política*, no. 31: 153–176.

Barciela López, C., and I. López Ortiz. 2000. "La Política de Colonización del Franquismo: Un Complemento de la Política de Riegos." In *El Agua en la Historia de España*, ed. C. Barcielo López and J. Melgarejo Moreno, 325–368. Alicante: Universidad de Alicante, Servicio de Publicaciones.

Barciela López, C., and I. López Ortiz. 2003. "El Fracaso de la Política Agraria del Primer Franquismo, 1939–1959. Vente Años Perdidos para la Agricultura Española." In *Autarquía y Mercado Negro*. In *El Fracaso Económico del Primer Franquismo, 1939–1959*, ed. C. Barciela López, 55–93. Barcelona: Crítica.

Barciela López, C., I. López Ortiz, J. Melgarefo Moreno, and J. A. Miranda Encarnación. 2001. *La España de Franco (1939–1975) Economía*. Madrid: Editorial Síntesis.

Baroja, P. [1909–1910] 1965. *Cesar o Nada*. Barcelona: Editorial Planeta.

Barry, M. 2013. *Material Politics: Disputes Along the Pipeline*. Oxford: Wiley-Blackwell.

Bauer, C. J. 1998. *Against the Current: Privatization, Water Markets, and the State in Chile*. New York: Springer Science+Business Media.

Beaumont, M. J., J. L. Beaumont, P. Arrojo, and E. Bernal. 1997. *El Embalse de Itoiz, la Razón o el Poder*. Bilbao: Bakeaz.

Becerril Bustamante, J. A. 2008. "Siglo y Medio de Innovaciones en Construcción: la Ingeniería Civil Española a través de la 'Revista de Obras Públicas.'" *Informes de la Construcción* 60 (510): 7–34.

Beltrán Villalva, M. 1996. "La Administración." In *La Época de Franco (1939–1975)*, Seria Historia de España Menendez Pidal, ed. R. Carr, 557–637. Madrid: Espasa Calpe.

Ben-Ami, S. 1983. *La Dictadura de Primo de Rivera (1923–1930)*. Barcelona: Editorial Planeta.

Bennett, J. 2010. *Vibrant Matter: A Political Ecology of Things*. Durham, NC: Duke University Press.

Benton, T. 1989. "Marxism and Natural Limits: An Ecological Critique and Reconstruction." *New Left Review* 178: 51–86.

Benton, T., ed. 1996. *The Greening of Marxism*. New York: Guilford Press.

Berman, M. 1982. *All That Is Solid Melts into Air: The Experience of Modernity*. New York: Simon and Schuster.

Bernal, A. M. 1990. "Agua para los Latifundios Andaluces." In *Agua y Modo de Produccíon*, ed. T. Pérez Picazo and G. Lemeunier, 271–310. Barcelona: Ed. Crítica.

Bernal, A. M. 2004. "Los Beneficiarios del Canal: Latifundios de Regadío." In *El Canal de los Presos (1940–1962). Trabajos Forzados: De la Represión Política a la Explotación Económica*, xxvii–xxxvi. Barcelona: Crítica.

Birch, K., L. Levidow, and T. Papaioannou. 2010. "Sustainable Capital? The Neoliberalization of Nature and Knowledge in the European 'Knowledge-based Bioeconomy.'" *Sustainability* 2 (9): 2898–2918.

Blackbourn D. 2006. *The Conquest of Nature. Water, Landscape and the Making of Modern Germany*. New York: W. W. Norton & Co.

Blanco Rodríguez, B., and A. Rodríguez González. 1999. "Sistemas de Financiación de las Obras Hidráulicas: Las Sociedades Estatales de Aguas." *Revista de Obras Públicas* 146 (3392): 7–30.

Boelens, R., D. Getsches, and A. Guevara-Gil, eds. 2010. *Out of the Mainstream: Water Rights, Politics and Identity*. London: Routledge.

Boelens, R., and N. C. Post Uiterweer. 2013. "Hydraulic Heroes: The Ironies of Utopian Hydraulism and Its Politics of Autonomy in the Guadalhorce Valley, Spain." *Journal of Historical Geography* 41: 44–58.

Bolton, W. 1954. *The Silver Spade: The Conrad Hilton Story*. New York: Farrar, Straus and Young.

Borrell Fontelles, J. 1993. "El Plan Hidrológico Nacional." *Revista de Obras Públicas* 140 (3321): 5.

Boyé, H. 2008. *Water, Energy, Desalination & Climate Change in the Mediterranean*. Sophia Antipolis, France: Blue Plan Regional Activity Center.

Braun, B., and N. Castree, eds. 1998. *Remaking Reality: Nature at the Millennium*. London: Routledge.

Brenan, G. 1943. *The Spanish Labyrinth*. Cambridge: Cambridge University Press.

Briggs, P. J. 1994. *Making American Foreign Policy: President-Congress Relations from the Second War to the Post Cold War Era*. Lanham, MD: Rowman & Littlefield.

Brunhes, J. 1920. *Geographie Humaine de la France*. Paris: Hanotaux.

Budds, J. 2004. "Power, Nature and Neoliberalism: The Political Ecology of Water in Chile." *Singapore Journal of Tropical Geography* 25 (3): 322–342.

Buesa, M. 1986. "Política Industrial y Desarrollo del Sector Eléctrico en España (1940–1963)." *Información Comercial Española* (June): 121–135.

Bullón, E. 1941. "Reformas Urgentes en la Enseñanza de la Geografía." *Estudios Geográficos*, no. 5: 661–678.

Butt, J. 1980. "The "Generation of 98": A Critical Fallacy." *Forum for Modern Language Studies* 16: 136–153.

Byrnes, M. S. 1999. ""Overruled and Worn Down": Truman Sends an Ambassador to Spain." *Presidential Studies Quarterly* 29 (2): 263–279.

Calvo, O. 1998. The Impact of American Aid in the Spanish Economy in the 1950s. MSc dissertation, Economic History, London School of Economics, London.

Calvo, O. 2001. "Bienvenido, Mister Marshall! La Ayuda Económica Americana y la Economía Española en la Década de 1950." *Revista de Historia Económica* 19 (Special Issue): 253–275.

Camprubí, L. 2012. ""Frankie the Frog": The Total Transformation of a River Basin as "Totalitarian" Technology (Spain, 1946–1961)." *Endeavour* 36 (1): 23–31.

Camprubí, L. 2014. Engineers and the Making of the Francoist Regime. Cambridge, MA: MIT Press.

Cano García, G. 1992. "Confederaciones Hidrográficas." In *Hitos Históricos de los Regadíos Españoles*, ed. A. Gil Olcina and A. Morales Gil, 309–334. Madrid: Ministerio de Agricultura, Pesca y Alimentación.

Capel, H. 1976. "La Geografía Española tras la Guerra Civil." *Geo Crítica*, no. 1: 5–35.

Caprotti, F. 2004. Nature and the City: Spectacle and Film Propaganda in New Towns in Fascist Italy, 1930–1939. D.Phil., School of Geography and the Environment, University of Oxford.

Caprotti, F. 2005a. "Information Management and Fascist Identity: Newsreels in Fascist Italy." *Media History* 11 (3): 177–191.

Caprotti, F. 2005b. "Fascism between Spectacle and Ideology." *Fast Capitalism* 1 (2). http://www.uta.edu/huma/agger/fastcapitalism/1_2/caprotti.html (accessed January 10, 2011).

Caprotti, F. 2007 *Mussolini's Cities: Internal Colonialism in Italy, 1930–1939*. Amherst, NY: Cambria Press.

Caprotti, F., and M. Kaika. 2008. "City and Nature: Ideology and Representation in Fascist New Towns." *Social and Cultural Geography* 9 (6): 613–634.

Cardenal, C. 1896. "¿Debe el Estado Construir Canales y Pantanos?" *Revista de Obras Públicas* 43 (22): 289–293.

Cardenal, C. 1901. "Servicio Hidrológico." *Revista de Obras Publicas* 48 (1339): 189–190.

Carr, R. 1995. *España: De la Restauración a la Democracia, 1875–1980.* Barcelona: Editorial Ariel, S.A.

Carr, R. 2001. *Modern Spain, 1875–1980.* Oxford: Oxford University Press.

Casada da Rocha, A., and J. A. Pérez. 1996. *Itoiz: Del Deber de la Desobediencia Civil al Ecosabotaje.* Pamplona: Pamiela.

Casado de Otaola, S. 2010. *Naturaleza Patria: Ciencia y Sentimiento de la Naturaleza en la España del Regeneracionismo.* Madrid: Fundación Jorge Juan/Marcial Pons Historia.

Casals Costa, V. 1988. "Defensa y Ordenacion del Bosque en España: Ciencia, Naturaleza y Sociedad en la Obra de los Ingenieros de Montes durante el Siglo XIX." *Geo Crítica,* no. 73: 5–63.

Casals Costa, V. 1996. *Los Ingenieros de Montes en la España Comtemporánea, 1848–1936.* Madrid: Ediciones del Serbal.

Casares, F. 1961. "Veinte Años de Política Hidráulica." *Revista de Obras Públicas* (December): 911.

Castillo-Puche, J. L. 1998. *Azorín y Baroja.* Madrid: Editorial Biblioteca Nueva, S.L.

Castoriadis, C. 1998. *The Imaginary Institution of Society.* Cambridge, MA: MIT Press.

Castro Valdiva, J. P. 2007. Comparecía de Don Juan Patricio Castro Valdiva, Vicerrector de Economía e Infraestructuras de la Universidad Politécnica de Cartagena. Murcia: Asamblea Regional de Murcia, Comisión Especial de Estudio sobre El Pacto del Agua.

Catalan, J. 2003. "La Reconstrucción Franquista y la Experiencia de la Europa Occidental, 1934–1959." In *Autarquía y Mercado Negro. El Fracaso Económico del Primer Franquismo, 1939–1959,* ed. C. Barciela López, 123–168. Barcelona: Crítica.

Cavestany y de Anduaga, R. 1958. *Una Política Agraria (Discursos).* Madrid: Dirección General de Coordinación, Crédito y Capitación del Ministerio de Agricultura.

Cazorla, A. 2000. "Early Francoism, 1939–1957." In *Spanish History since 1808,* ed. J. Alvarez Junco and A. Shubert, 259–276. London: E. Arnold.

CEC. 2000. Directive 2000/60/EC of the European Parliament and of the Council of 23 October 2000—Establishing a Framework for Community Action in the Field of Water Policy. Brussels: Official Journal of the Commission of the European Communities.

References

Centro de Estudios y Experimentación de Obras Públicas. 2011. Evaluación del Impacto del Cambio Climático en los Recursos Hídricos en Régimen Natural. Resumen Ejecutivo. Madrid: Centro de Estudios y Experimentación de Obras Públicas (CEDEX), Dirección General del Agua & Oficina Española de Cambio Climático (OECC).

Chari, R., and P. M. Heywood. 2009. "Analysing the Policy Process in Democratic Spain." *West European Politics* 32 (1): 26–54.

Cheyne, G. J. G. 1972. *Joaquín Costa, el gran Desconocido*. Barcelona: Ediciones Ariel.

Cimadevilla Costa, C., and J. A. Herreras Espino. 1995. "La Urgente Necesidad del Plan Hidrológico Nacional." *Revista de Obras Públicas* 142 (3345): 7–14.

Conca, K. 2005. *Governing Water: Contentious Transnational Politics and Global Institution Building*. Cambridge, MA: MIT Press.

Conde de Guadalhorce. 1950. "Acción Colectiva para el Desarrollo Hidroeléctrico." *Revista de Obras Públicas* (October): 504–510.

Consejo Económico y Social. 1996. "Sobre Recursos Hídricos en España. Incidencia en el Sector Agrario." In *Colección Informes 1/1996*. Madrid: Consejo Económico y Social.

Cook, I., and E. Swyngedouw. 2012. "Cities, Social Cohesion and the Environment: Towards a Future Research Agenda." *Urban Studies* 49: 1938–1958.

Coordinadora de Organizaciones de Agricultores y Ganaderos. 1993. "Alegaciones al Borrador del Plan Hidrológico Nacional," October, typescript, 21 pp. Madrid: COAG.

Coordinadora de Organizaciones de Agricultores y Ganaderos. 1994. "Observaciones de COAG al Informe sobre las Propuestas de Modificación del Anteproyecto del Plan Hidrológico Nacional." Madrid: COAG.

Coordinadora de Organizaciones de Defensa Ambiental. 1994. Consideraciones al Nuevo Anteproyecto de Ley del Plan Hidrológico Nacional. Madrid: CODA.

Costa, J. [1892] 1975. *Política Hidráulica y Misión Social de los Riegos en España*. Madrid: Edición de la Gaya Ciencia.

Costa, J. [1900] 1981. *Reconstitución y Europeización de España*. Madrid: Instituto de Estudios de Administración Local.

Costa, J. [1901] 1998. *Oligarquía y Caciquismo*. Madrid: Editorial Biblioteca Nueva.

Costa, J. 1912. El Arbolada y la Patria. Madrid: Biblioteca J. Costa.

Costa Martínez, J. 1911. *Política Hidráulica (Misión social de los Riegos en España)*. Madrid: Biblioteca J. Costa.

Couchoud Sebastia, R. 1977. "Análisis Crítico de la Ley de Aguas y Legislación de Obras Hidráulicas en España." *Revista de Obras Públicas* 124 (3144): 247–258.

Cronon, W. 1991. *Nature's Metropolis: Chicago and the Great West*. New York: W. W. Norton.

Cuerpo de Ingenieros. 1899a. *Avance de un Plan General de Pantanos y Canales de Riego*. Madrid: Cuerpo de Ingenieros de Caminos, Canales y Puertos.

Cuerpo de Ingenieros. 1899b. "Consejo de Ministros Presido por S. M. La Reina." *Revista de Obras Públicas* 46 (1233): 191–192.

Cuerpo de Ingenieros. 1899c. "Junta de Representación del Cuerpo de Ingenieros de Caminos, Canales y Puertos Presidida por Práxedes Mateo Sagasta." *Revista de Obras Públicas* 46 (1235): 219–220.

Cuerpo de Ingenieros. 1899d. "Pantanos y Canales de Riego." *Revista de Obras Públicas* 46 (1230): 143–147.

Cuerpo de Ingenieros. 1899e. "Pantanos y Canales de Riego." *Revista de Obras Públicas* 46 (1229): 131–134.

de Cervantes López, M. 1865. "Importancia de los Estudios que han de verificar las Divisiones Hidrológicas." *Revista de Obras Públicas* 13: 262–264.

de Maeztu, R. [1899] 1997. *Hacia otra España*. Madrid: Editorial Biblioteca Nueva.

de Palau Catalá, M. 1875. "Divisiones Hidrológicas." *Revista de Obras Públicas* 23: 234–238.

de Reparez, G. 1906. "Hidráulica y Dasonomía." *Diario de Barcelona*, July 21, 1906.

de Riquer, B. 2010. *La Dictadura de Franco*. Barcelona: Crítica/Marcial Pons.

de Unanumo, M. [1902] 1998. *El Torno de Casticismo*. Madrid: Biblioteca Nueva.

de Unamuno, M. 1968. *Obras Completas*. Madrid: Edición Escélicer.

del Campo y Francés, A. 1992. *José Torán—Un Ingeniero Insólito*. Madrid: Colegio de Ingenieros de Caminos, Canales y Puertos.

Del Corral, E. 1959a. "Cuatro Mil Kilómetros recorrió Franco por Tierras de Aragón y Cataluña." *Diario ABC*, July 6, 1959, 1.

Del Corral, E. 1959b. "Un Jornada de Acción Política—Emoción Auténtica de los Labradores." *Diario ABC*, July 1, 1959, 1–2.

del Moral Ituarte, L. 1991. *La Obra Hidráulica en la Cuenca Baja del Guadalquivir (Siglos XVIII–XX)—Gestión del Agua y Organización del Territorio*. Sevilla: Consejería de Obras Públicas y Transportes, Junta de Andalucía/Secretario de Publicaciones, Universidad de Sevilla.

del Moral Ituarte, L. 1994. "Interbasin Hydraulic Transfers in the Spanish Water Management." *Etudes Vauclusiennes*, 91–94.

del Moral Ituarte, L. 1995. "El Origen de la Organización Administrativa del Agua y de los Estudios Hidrológicos en España. El Caso de la Cuenca del Guadalquivir." *Estudios Geográficos* 56 (219): 371–393.

del Moral Ituarte, L. 1996. "Sequía y Crisis de Sostenibilidad del Modelo de Gestión Hidráulica." In *Clima y Agua—La Gestión de un Recurso Climático*, ed. M. V. Marzol, P. Dorta, and P. Valladares, 179–187. Madrid: Tabapress.

del Moral Ituarte, L. 1998. "L'Etat de la Politique Hydraulique en Espagne." *Hérodote*, no. 91: 118–138.

del Moral Ituarte, L. 1999. "La Política Hidráulica en España de 1936 a 1996." In *El Agua en los Sistemas Agrarios*. In *Una Perspectiva Histórica*, ed. R. Garrabou and J. M. Naredo, 181–195. Madrid: Fundación Argentaria.

del Moral Ituarte, L. 2005. *La Gestión del Agua en Andalucía—Aspectos Económicos, Políticos y Territoriales*. Sevilla: Mergablum, S.L.

del Moral Ituarte, L. 2008. *Análisis de los Discursos Relacionados con los Procesos Territoriales que Condicionan la Presión sobre el Agua en la Cuenca del Segura*. Madrid: Colegio de Geógrafos.

del Moral Ituarte, L. 2009. "Nuevas Tendencias en Gestión del Agua, Ordenación del Territorio e Integración de Políticas Sectoriales." In *Scripta Nova—Revista Electrónica de Geografía y Ciencias Sociales*. Barcelona: Universidad de Barcelona.

del Moral Ituarte, L., P. van der Werff, K. Bakker, and J. Handmer. 2003. "Global Trends in Water Policy in Spain." *Water International* 28 (3): 358–366.

del Prado y Palacio, J. 1917. *Hagamos Patria. Estudio Político y Económico de Problemas Nacionales de Inaplazable Solución*. Madrid: Ed. Tipografía Artística.

del Rio Cisneros, A. 1964. *Pensamiento Político de Franco—Antología*. Madrid: Servicio Informativo Español.

Deleuze, G., and F. Guattari. 1983. *Anti-Oedipus*. Minneapolis: Minnesota University Press.

Delli Priscoli, J., and A. T. Wolf. 2009. *Managing and Transforming Water Conflicts*. Cambridge: Cambridge University Press.

Díaz-Marta, M. 1993. "Notas para un Análisis del Plan Hidrológico en Preparación." *Revista de Obras Públicas* 140 (3318): 7–16.

Díaz-Marta, M. 1996. "Creatividad, estancamiento y evolución de la política del agua en España." *OP: Revista del Colegio de Ingenieros de Caminos, Canales y Puertos*, no. 37: 4–13.

Díaz-Marta Pinilla, M. [1969] 1997. *Las Obras Hidráulicas en España*. Aranjuez: Ediciones Doce Calles, S. L.

di Lampedusa, G. 1960. *The Leopard*. New York: Pantheon Books.

Dirección General de Obras Hidráulicas. 1990. *Plan Hidrológico—Síntesis de la Documentación Básica*. Madrid: Ministerio de Obras Públicas y Urbanismo.

Director General de Obras Hidráulicas. 1971. Discursos Pronunciados en el Acto Conmemorativo de la Fundación del Centro de Estudios Hidrográficos. *Revista de Obras Públicas* (May): 396–397.

Douglas, I. R. 1997. "The Calm before the Storm: Virilio's Debt to Foucault, and Some Notes on Contemporary Global Capital." http://nideffer.net/proj/_SPEED_/1.4/articles/douglas.html (accessed October 20, 2014).

Downward, S. R., and R. Taylor. 2007. "An Assessment of Spain's Programa AGUA and Its Implications for Sustainable Water Management in the Province of Almería, Southeast Spain." *Journal of Environmental Management* 82: 277–289.

Driever, S. L. 1998a. ""And Since Heaven Has Filled Spain with Goods and Gifts": Lucas Mallada, the Regeneracionist Movement, and the Spanish Environment, 1881–90." *Journal of Historical Geography* 24: 36–52.

Driever, S. L. 1998b. "Mallada y el Regeneracionismo Español." In *Lucas Mallada—La Futura Revolución Española y otros Escritos Regeneracionistas*, ed. F. J. Ayala-Carcedo and S. L. Driever, 15–61. Madrid: Biblioteca Nueva.

Duffy, M. 2011. Innovations Solutions within the Water Industry: Desalination. American Water. http://www.amwater.com/files/InnovationsSolutionsWithinTheWaterIndustryDesalination.pdf (accessed October 20, 2014)

Dunlap T. 1999. *Nature and the English Diaspora: Environment and History in the United States, Canada, Australia, and New Zealand*. Cambridge: Cambridge University Press.

Edwards, J. 1999. *Anglo-American Relations and the Franco Question, 1945–1955*. Oxford: Oxford University Press.

Ekbladh, D. 2002. ""Mr. TVA": Grass-Roots Development, David Lilienthal, and the Rise and Fall of the Tennessee Valley Authority as a Symbol for U.S. Overseas Development, 1933–1973." *Diplomatic History* 26 (3): 335–374.

Elimelech, M., and W. A. Phillip. 2011. "The Future of Seawater Desalination: Energy, Technology, and the Environment." *Science* 333: 712–717.

Eliot, T. S. [1922] 1969. *"The Waste Land."* In *The Complete Poems and Plays*, 59–80. London: Faber and Faber.

Embid, A. 1998. *El Nuevo Derecho del Aguas: Las Obras Hidráulicas y su Financialicación*. Zaragoza: Seminario del Derecho de Agua de la Universidad de Zaragoza, Confederación Hidrográfica del Ebro.

Embid, A. 2002. "The Evolution of Water Law and Policy in Spain." *International Journal of Water Resources Development* 18 (2): 261–283.

References

Escartin Hernández, C. M., F. Cabezas Calvo-Rubio, and F. Estrada Lorenzo. 1999. "La Política del Agua." *Revista de Obras Públicas*, no. 3388: 79–85.

Escobar, A. 1999. "After Nature: Steps to an Anti-essentialist Political Ecology." *Current Anthropology* 40: 1–30.

European Environment Agency. 2007. *Climate Change and Water Adaptation Issues*. EEA Technical Report 2/2007. Copenhagen: European Environment Agency.

Fanlo Loras, A. 1996. *Las Confederaciones Hidrográficas y Otras Administraciones Hidráulicas*. Madrid: Editorial Civitas.

Fernández Almagro, M. 1970. *Historia Política de la España Contemporánea Vol. 3. 1897–1902*. 2nd ed. Madrid: Editorial Alianza.

Fernández Clemente, E. 1986. "Las Confederaciones Sindicales Hidrográficas durante la Dictadura de Primo de Rivera: La C.H.S. del Ebro." In *La Hacienda Pública en la Dictadura (1923–1930)*, ed. J. Velarde Fuertes, 335–361. Madrid: Instituto de Estudios Fiscales.

Fernández Clemente, E. 1990. "La Política Hidráulica de Joaquín Costa." In *Agua y Modo de Producción*, ed. T. Pérez Picazo and G. Lemeunier, 69–97. Barcelona: Editorial Crítica.

Fernández Clemente, E. 2000. *Un Siglo de Obras Hidráulicas en España: De la Utopía de Joaquín Costa a la Intervención del Estado*. Zaragoza: Facultad de Ciencias Económicas y Empresariales, Universidad de Zaragoza.

Fernández de Valderrama, G. 1964. "España-USA, 1953–1964." *Economía Financiero* 6: 14–51.

Figuero, J. 1998. *La España de la Rabia y de la Idea*. Barcelona: Plaza y Janés Editores.

Figuero, J., and C. G. Santa Cecilia. 1998. *La España del Desastre*. Barcelona: Plaza y Janés Editores.

Figuerola Paloma, M. 1999. "La Transformación del Turismo en un Fenómeno de Masas. La Planificación Indicativa (1950–1974)." In *Historia de la Economía del Turismo en España*, ed. C. Pellejero, 77–134. Madrid: Civitas.

Fischer-Kowalski, M. 1998. "Society's Metabolism. The Intellectual History of Material Flow Analysis, Part I, 1860–1970." *Journal of Industrial Ecology* 2: 61–78.

Fischer-Kowalski, M. 2003. "On the History of Industrial Metabolism." In *Perspectives on Industrial Ecology*, ed. D. Bourg and S. Erkman, 33–45. Sheffield, UK: Greenleaf Publishing.

Font, N., and J. Subirats. 2010. "Water Management in Spain: The Role of Policy Entrepreneurs in Shaping Change." *Ecology and Society* 15 (2). http://www.ecologyandsociety.org/vol15/iss2/art25/ (accessed Otober 20, 2014).

Fontana, J. 1975. *Cambio Económico y Actitudes Políticas en la España del Siglo XIX*. Barcelona: Editorial Ariel.

Foro Ciudadano. 2005. El Nacionalismo Hidráulico. El Verdad http://www.forociudadano.org/index.php/opinion/200-el-nacionalismo-hidraulico-la-verdad-180705 (accessed October 20, 2014).

Foster, J. B. 2000. *Marx's Ecology: Materialism and Nature*. New York: Monthly Review Press.

Foucault, M. 2008. *The Birth of Biopolitics: Lectures at the Collège de France 1978–1979*. Basingstoke: Palgrave.

Franco, F. 1940. "Prólogo." In *Contribución al Estudio Estratégico de la Península: Geografía Militar de España. Paises y Mares Limítrofes*, ed. J. Díaz de Villegas, i–xxvii. Madrid: Servicio Geográfico y Cartográfico.

Franco, F. 1959a. "Inauguration of the Hydraulic Works in Lérida." *Diario ABC*, nr. 16630, July 1, 1.

Franco, F. 1959b. "Speech of Chief of State in Medina del Campo." *Diario ABC*, nr. 16734, October 30, 1.

Franco, F. 1959c. "Discourse of the Head of State, November 3, 1959." *Diario ABC*, nr. 16738, November 4, 1.

Franco, F. 1960. *Discursos y Mensajes del Jefe del Estado 1955–1959*. Madrid: Dirección General de Información. Publicaciones Españolas.

Franco, F. 1964. *Discursos y Mensajes del Jefe del Estado 1959–1963*. Madrid: Dirección General de Información. Publicaciones Españolas.

Frutos Mejias, L. 1995. "Las Confederaciones Sindicales Hidrográficas (1926–1931)." In *Planificación Hidráulica en España*, ed. A. Gil Olcina and A. Morales Gil, 181–255. Murcia: Fundación Caja del Mediterráneo (CAM).

Fundación Cajamar. 2009. *La Desalación en España*. Almería: CAJAMAR.

Furlong, K. 2010. "Neoliberal Water Management: Trends, Limitations, Reformulations." *Environment and Society: Advances in Research* 1: 46–75.

Fusi, J. P., and J. Palafox. 1989. *España 1808–1996: El Desafío de la Modernidad*. Madrid: Editorial Espasa.

Gadgil, M., and G. Ramachandra. 1992. *This Fissured Land: An Ecological History of India*. Delhi: Oxford University Press.

Gallo, M. 1974. *Spain under Franco: A History*. New York: E. P. Dutton and Co., Inc.

Gamir, L. 1999. *Las Privatizaciones en España*. Madrid: Ediciones Pirámide.

Gandy, M. 2002. *Concrete and Clay: Reworking Nature in New York City*. Cambridge, MA: MIT Press.

References

Gandy, M. 2005. "Cyborg Urbanization: Complexity and Monstrosity in the Contemporary City." *International Journal of Urban and Regional Research* 29: 26–49.

García Alonso, J., and J. Iranzo Martín. 1988. *La Energía en la Economía Mundial y en España*. Madrid: Editorial AC, Libros Científicos y Técnicos.

García Alvarez, J. 2002. *Provincias, Regiones y Comunidades Autónomas. La Formación del Mapa Político de España*. Madrid: Secretaría General del Senado, Dirección de Estudios y Documentación, Departamento de Publicaciones.

García Díez, J. A. 2001. *Ribadelago—Tragedia de Vega de Tera*. Zamora: A. Saveedra.

García Hernández, R. 1876. "Divisiones Hidrológicas." *Revista de Obras Públicas* 24: 172–173.

García Lozano, R. Á. 2009. "La Catástrofe de Ribadelago en la Prensa Nacional—Sábado Gráfico." *Brigecio: Revista de Estudios de Benavente y sus Tierras*, nos. 18–19: 307–312.

García Novo, F., J. Toja Santillana, and C. Granado-Lorencio. 2010. "The State of Water Ecosystems." In *Water Policy in Spain*, ed. A. Garrido and R. M. Llamas, 21–28. London: CRC Press.

García Yelo, J. J. 1997. "Situación Actual del Trasvase Tajo Segura." *Revista de Obras Públicas* (3365): 7–17.

Garrabou, R. 1975. "La Crisis Agrária Espanyola de Finals del Segle XIX: Una Etapa del Desenvolupament del Capitalisme." *Recerques* 5: 163–216.

Garrabou, R. 1997. "Políticas Agrarias y Desarrollo de la Agricultura Española Contemporánea: Unos Apuntos." *Papeles de Economía Española*, no. 73: 141–148.

Garrido, A. 2006. "Analysis of Spanish Water Law Reform." In *Water Rights Reform: Lessons for Institutional Design*, ed. B. R. Bruns, C. Ringler, and R. Meinzen-Dick, 219–236. Washington, DC: International Food Policy Research Institute.

Garrido Moyron, J. 1957. "Desarrollo Hidroeléctrico Español en La Producción de Energía desde 1939 a 1955 en Relación con los Recursos Hidroeléctricos Totales de España." *Revista de Obras Públicas* (December): 661–667.

Gasset, R. 1899. "Pantanos y Canales de Riego—Real Decreto—11 Mayo 1900." *Revista de Historia Económica* 47 (Special Issue—Supplement to Issue 1286): i–vii.

Gasset, R. 1916. *Reforma de los Presupuestos: El Pan Extraordinario de Obras Públicas*. Madrid: Ministerio de Fomento.

Giasanta, C. 1999. "In-depth Analysis of Relevant Stakeholders: Guadalquivir River Basin Authority." Mimeograph, 22 pp. Seville: Department of Geography, Universidad de Sevilla.

Giansante, C., L. Babiano, and L. del Moral Ituarte. 2000. "L'Evolution des Modalités d'Allocation des Ressources en Eau en Espagne." *Revue de l'Economie Méridionale* 48 (191): 235–247.

Giddens, A. 1991. *The Mental and the Material*. London: Verso.

Gil Olcina, A. 1992. "Desequilibrios Hidrográficos en España y Trasvases a la Vertiente Mediterránea: Utopías y Realizaciones." *Investigaciones Geográficos*, no. 10: 7–23.

Gil Olcina, A. 1995. "Conflictos Autonómicos sobre Trasvases de Agua en España." *Investigaciones Geográficas* (13): 17–28.

Gil Olcina, A. 2001. "Del Plan General de 1902 a la Planificación Hidrológica." *Investigaciones Geográficas* 25: 5–31.

Gil Olcina, A. 2002. "De Los Planos Hidráulicos a la Planificación Hidrológica." In *Insuficiencias Hídricas y Plan Hidrológico Nacional*, ed. A. Gil Olcina and A. Morales Gil, 11–44. Alicante: Caja de Ahorro del Mediterráneo/Instituto Universitario de Geografía.

Gil Olcina, A. 2003. "Perduración de los Planes Hidráulicos en España." In *La Directiva Marco del Agua: Realidades y Futuros*, ed. P. Arrojo Agudo and L. del Moral Ituarte, 29–61. Zaragoza: Fundación Nueva Cultura del Agua, Institución Fernando el Católico, Universidad de Zaragoza.

Gil Olcina, A. 2010. "De los Planes Hidráulicos a la Planificación Hidrológico." Instituto Universitaria de Geografía, Universidad de Alicante, December 1. http://www.cervantesvirtual.com/portales/investigaciones_geograficas/partes/325899/n-25--2001 (accessed December 1, 2010).

Gil Olcina, A., and A. M. Rico Amorós. 2008. *Políticas del Agua III—De la Ley de Aguas de 1986 al PHN*. Murcia: ESAMUR.

Gil Olcina, A., and A. Morales Gil, eds. 1992. *Hitos Históricos de los Regadíos Españoles*. Madrid: Ministerio de Agricultura, Pesca y Alimentación.

Giménez Pérez, F. 2002. "Comparación del diagnóstico de Joaquín Costa acerca de la España de 1899 con la situación de España en 1999." *El Catobeplas—Revista Crítica del Presente* 6: 24.

Global Water Intelligence. 2010. *IDA Desalination Yearbook 2010–2011*. Oxford/Topsfield, MA: Global Water Intelligence/International Desalination Association.

Godelier, M. 1986. *The Mental and the Material*. London: Verso.

Gomez De Pablos, M. 1972. "Principios Rectores de una Política Hidráulica." *Revista de Obras Públicas* 119 (3086): 463–475.

Gomez De Pablos, M. 1973a. "El Centro de Estudios Hidrográficos y la Planificación Hidráulica Española." *Revista de Obras Públicas* 120 (3096): 241–248.

References

Gomez De Pablos, M. 1973b. "El Desarrollo de los Recursos Hidráulicos en España." *Revista de Obras Públicas*: 337–344.

Gómez Mendoza, J. 1989. "La Discusión Técnica en Torno a la Política Hidráulica y la Política Forestal antes del Plan Nacional de Obras Hidráulicas." In *Los Paisajes del Agua*, 85–96. Libro Jubilar dedicado al Profesor Antonio López Gómez, Universitat de Valéncia/Universidad de Alicante.

Gómez Mendoza, J. 1992a. "Regeneracionismo y Regadíos." In *Hitos Históricos de los Regadíos Españoles*, ed. A. Gil Olcina and A. Morales Gil, 231–262. Madrid: Ministerio de Agricultura, Pesca y Alimentación.

Gómez Mendoza, J. 1992b. *Ciencia y Política de los Montes Españoles (1848–1936)*. Madrid: Icona.

Gómez Mendoza, J. 1997. "La Formación de la Escuela Española de Geografía (1940–1952)." *Ería* (42): 107–146.

Gómez Mendoza, J., and L. del Moral Ituarte. 1995. "El Plan Hidrológico Nacional: Criterios y Directrices." In *Planificación Hidráulica en España*, ed. A. Gil Olcina and A. Morales Gil, 331–378. Murcia: Fundación Caja del Mediterráneo.

Gómez Mendoza, J., and R. Mata Olma. 2002. "Reploblación Forestal y Territorio (1940–1971). Marco Doctrinal y Estudio de la Sierra de los Filabres (Almería)." *Ería* 58: 129–155.

Gómez Mendoza, J., and N. Ortega Cantero. 1987. "Geografía y Regeneracionismo en España." *Sistema* 77: 77–89.

Gómez Mendoza, J., and N. Ortega Cantero. 1992. "Interplay of State and Local Concern in the Management of Natural Resources: Hydraulics and Forestry in Spain 1855–1936." *GeoJournal* 26: 173–179.

Gómez-Baggethun, E., R. de Groot, P. L. Lomas, and C. Montes. 2010. "The History of Ecosystem Services in Economic Theory and Practice: From Early Notions to Markets and Payment Schemes." *Ecological Economics* 69 (6): 1209–1218.

González Bernáldez, F. 1981. *Ecología y Paisaje*. Madrid: H. Blume Ediciones.

González Bernáldez, F. 1989. "Ecosistema Áridos y Endorreicos Españoles." In *Zonas Áridas en España*, ed. Real Académica de Ciencias Exactas, Físicas y Naturales, 223–238. Madrid: Real Académica de Ciencias Exactas, Físicas y Naturales.

González Martín, F. J. 2004. "Filosofía del Derecho y Regeracionismo Político en el Concepto de Estado de Joquín Costa." Doctoral dissertation, Faculty of Law, Complutense University, Madrid.

González Paz, J. 1970. "La Significación del Transvase en Nuestra Política Hidráulica." *Revista de Obras Públicas* (October): 983–992.

González Paz, J. 1994. "Reflexiones Personales en Torno al Problema de los Trasvases." *Revista de Obras Públicas* 141 (3337): 33–42.

González Quijano, P. M. 1913. "La Política Hidráulica en España—Apuntes de Psicología Coleectiva." *Revista de Obras Públicas* 61 (1983): 473–476.

Gonzalo Fernández de la Mora, D. 1971. "Discurso de D. Gonzalo Fernández de la Mora, Ministro de Obras Publicas." *Revista de Obras Públicas* (May): 398–399.

Grindlay, A. L., M. Zamorano, M. I. Rodríguez, E. Molero, and M. A. Urrea. 2011. "Implementation of the European Water Framework Directive: Integration of Hydrological and Regional Planning at the Segura River Basin, Southeast Spain." *Land Use Policy* 28: 242–256.

Grindley Moreno, A., and E. Hernández Gómez-Arboleya. 2010. *Las Infraestructuras Hidráulicas en la Cuenca del Segura*. Madrid: Colegio del Ingenieros de Caminos, Canales Y Puertos.Grundman, R. 1991. *Marxism and Ecology*. Oxford: Clarendon Press.

Guàrdia, M., M. Rosselló, and S. Garriga. 2014. "Barcelona's Water Supply, 1867–1967: The Transition to a Modern System." *Urban History* 41 (3): 415–434.

Guerra del Río, R. 1933. "Política Hidráulica—Plan Nacional." In *Plan Nacional de Obras Hidráulicas*, ed. M. L. Pardo, 3–8. Madrid: Centro de Estudios Hidrográficas.

Guirao, F. 1998. *Spain and the Reconstruction of Western Europe, 1945–1957. Challenge and Response*. Oxford: Macmillan Press.

Gutiérrez Molina, J. L. 2004. "Por Soñar con la Libertad, los Convirtieron en Esclavos. Presos, Prisioneros y Obras Públicas y Privadas en Andalucía durante la Guerra Civil." *Historia Actual Online*, no. 3 (Winter 2004): 39–54.

Gutiérrez Molina, J. L. 2006. "Franquismo, Latifundistas y Obras Hidráulicas en Andalucía: El Canal de los Presos." *Cuadernos para el Diálogo*, no. 14: 16–23.

Haddad, B. 2000. *Rivers of Gold: Designing Markets to Allocate Water in California*. Washington, DC: Island Press.

Haraway, D. 1991. *Simians, Cyborgs and Women: The Reinvention of Nature*. London: Free Association Books.

Haraway, D. 1997. *Modest_Witness@Second_Millenniun.FemaleMan©_Meets_Onco-Mouse^{tm}*. London: Routledge.

Harris, L. M. 2002. "Water and Conflict Geographies of the Southeastern Anatolia Project." *Society and Natural Resources* 15: 743–759.

Harris L. M. 2012. "State as Socionatural Effect: Variable and Emergent Geographies of the State in Southeastern Turkey." *Comparative Studies of South Asia, Africa and the Middle East* 32 (1): 25–39.

References

Harris, L. M., and S. Alatout. 2010. "Negotiating Hydro-Scales, Forging States: Comparison of the Upper Tigris/Euphrates and Jordan River Basins." *Political Geography* 29 (3): 148–156.

Harrison, J. 1973. "The Spanish Famine of 1904–6." *Agricultural History* 47 (4): 300–307.

Harrison, J. 2000a. "Introduction: The Historical Background to the Crisis of 1898." In *Spain's 1898 Crisis: Regenerationism, Modernism, Post-colonialism*, ed. J. Harrison and A. Hoyle, 1–8. Manchester: Manchester University Press.

Harrison, J. 2000b. "Tackling National Decadence: Economic Regenerationism in Spain after the Colonial Debacle." In *Spain's 1898 Crisis: Regenerationism, Modernism, Post-colonialism*, ed. J. Harrison and A. Hoyle, 55–80. Manchester: Manchester University Press.

Harrison, J., and A. Hoyle, eds. 2000. *Spain's 1898 Crisis: Regenerationism, Modernism, Post-Colonialism*. Manchester: Manchester University Press.

Harvey, A. D. 1999. "The Body Politic: Anatomy of a Metaphor." *Contemporary Review* 275: 85–93.

Harvey, D. 1982. *The Limits to Capital*. Oxford: Blackwell.

Harvey, D. 1989. *The Condition of Postmodernity*. Oxford: Blackwell.

Harvey, D. 1996. *Justice, Nature, and the Geography of Difference*. Oxford: Blackwell.

Henn, D. 2003. "History, Philosophy, and Fiction: Pío Baroja's *César o Nada*." *Neophilologus* 87: 233–246.

Hernández, J. M. 1994. "La Planificación Hidrológica en España." *Revista de Estudios Agro-Sociales*, no. 167: 13–25.

Heynen, N., M. Kaika, and E. Swyngedouw. 2006. *In the Nature of Cities: Urban Political Ecology and the Politics of Urban Metabolism*. London: Routledge.

Heynen, N., J. McCarthy, S. Prudham, and P. Robbins, eds. 2007. *Neoliberal Environments: False Promises and Unnatural Consequences*. London: Routledge.

Holifield, R., M. Porter, and G. Walker, eds. 2010. *Spaces of Environmental Justice*. Oxford: Wiley-Blackwell.

Holman, O. 1996. *Integrating Southern Europe: EC Expansion and the Transnationalization of Spain*. London: Routledge.

Hoyle, A. 2000a. "The Function of Landscape in Baroja's *La Luch por la Vida*." In *Spain's 1898 Crisis: Regenerationism, Modernism, Post-colonialism*, ed. J. Harrison and A. Hoyle, 181–194. Manchester: Manchester University Press.

Hoyle, A. 2000b. "Introduction: The Intellectual Debate." In *Spain's 1898 Crisis: Regeneracionism, Modernism, Post-colonialism*, ed. J. Harrison and A. Hoyle, 9–51. Manchester: Manchester University Press.

Hughes, J. 2000. *Ecology and Historical Materialism*. Cambridge: Cambridge University Press.

ICEX. 2009. "Water Desalination in Spain—Water Cycle 02." Industry Reports 2009. Madrid: Instituto Español de Comercio Exerior, Gobierno de España.

ICEX. 2010. "New Technologies in Spain—Desalination." Advertising supplement to *MIT Technology Review*.

ICEX. n.d. "Water Treatment & Desalination—Spain." Spanish Institute for Foreign Trade/Trade Commission of Spain, August 1. http://www2.technologyreview.com/microsites/spain/water/docs/Spain_desalination.pdf (accessed August 1, 2011)

Iglesias, A., M. Moneo, L. Garrote, and F. Flores. 2010. "Drought and Climate Risk." In *Water Policy in Spain*, ed. A. Garrido and R. M. Llamas, 63–75. London: CRC Press.

İlhan, A. 2009. "Social Movements in Sustainability Transitions: Identity, Social Learning, and Power in the Spanish and Turkish Water Domains." PhD diss., ICTA, Autonomous University of Barcelona, Bellaterra, Barcelona.

Isern, D. 1899. *El Desastre Nacional y sus Causas*. Madrid: Imprenta de la Viuda de Minuesa de los Ríos.

Jacobs N. J. 2003. *Environment, Power, and Injustice: A South African History*. Cambridge: Cambridge University Press.

Jáuregui, P., and A. M. Ruiz-Jiminéz. 2005. "A European Spain: The Recovery of Spanish Self-Esteem and International Prestige." In *Entangled Identities: Nations and Europe*, ed. A. Ichigo and W. Spohn, 72–87. Aldershot: Ashgate.

Jefatura del Estado. 2011. "Ley 10/2001, de 5 de Julio, del Plan Hidrológico Nacional." *Bulletin Official del Estado* 161 (July 6, 2001): 24228–24250.

John Paul II. 2005. "Discurso del Santo Padre Juan Pablo II a los Obispos Españoles en Vista 'Ad Limina,'" Lunes 24 de enero de 2005, January 25, 2005. http://www.vatican.va/holy_father/john_paul_ii/speeches/2005/january/documents/hf_jp-ii_spe_20050124_spanish-bishops_sp.html (accessed December 21, 2011).

Kaika, M. 2003. "The WFD: A New Directive for a Changing Social, Political and Economic European Framework." *European Planning Studies* 11 (3): 299–316.

Kaika, M. 2004. "Water for Europe: The Creation of the European Water Framework Directive." In *Managing Water Resources: Past and Present*, ed. J. Trottier and P. Slack, 89–116. Oxford: Oxford University Press.

Kaika, M. 2005. *City of Flows: Modernity, Nature and the City*. London and New York: Routledge.

Kaika, M. 2006. "Dams as Symbols of Modernization: The Urbanization of Nature between Geographical Imagination and Materiality." *Annals of the American Association of Geographers* 96 (2): 276–301.

Kaika, M. 2010. "Architecture and Crisis: Re-inventing the Icon, Re-imag(in)ing London and Re-branding the City." *Transactions of the Institute of British Geographers* 35 (4): 453–474.

Kaika, M., and B. Page. 2003. "The EU Water Framework Directive: Part 1. European Policy-making and the Changing Topography of Lobbying." *Environmenal Policy and Governance* 13 (6): 314–327.

Kaika, M., and E. Swyngedouw. 1999. "Fetishising the Modern City: The Phantasmagoria of Urban Technological Networks." *International Journal of Urban and Regional Research* 24: 120–138.

Kallis, G. 2011. "In Defence of Degrowth." *Ecological Economics*, no. 70: 873–880.

Lafuente, E. 1990. "Un Gran Proyecto Dormido." *Revista de Obras Públicas* 137 (3289): 45–49.

Lafuente, I. 2002. *Esclavos por la Patria. La Explotación de los Presos bajo el Franquismo*. Madrid: Editorial Temas de Hoy.

Latour, B. 1993. *We Have Never Been Modern*. London: Harvester Wheatsheaf.

Lefebvre, H. 1991. *The Production of Space*. Oxford: Blackwell.

Lera, J. 1999. "40 Años de la Tragedia de Ribadelago, en la que Murieron 144 Personas." *El País*, January 10.

Levins, R., and R. Lewontin. 1985. *The Dialectical Biologist*. Cambridge, MA: Harvard University Press.

Lewontin, R., and R. Levins. 2007. *Biology under the Influence*. New York: Monthly Review Press.

Liedtke, B. N. 1998. *Embracing a Dictatorship. US Relations with Spain, 1945–1953*. Oxford: St. Martin's Press/Macmillan.

Linton, J. 2010. *What Is Water? The History of a Modern Abstraction*. Vancouver: University of British Columbia Press.

Llamas Madurga, M. R. 1984. "Política Hidráulica y Génesis de Mitos Hidráulicos en España." *Cimbra*, no. 218: 16–25.

Llamas Madurga, M. R. 1993. "El Plan Hidrológico Nacional: Presentación Adrián Baltanás García en la Revista de Obras Públicas, Mayo 1993, 7–11." *Revista de Obras Públicas* 140 (3326): 99–101.

López Bermúdez, F. 1974. "El Trasvase Tajo-Segura." *Estudios Geográficos* 35 (135): 320–330.

López Ontiveros, A. 1992. "Significado, Contenido, Temática, Ideología de los Congresos Nacionales de Riegos (1913–1934)." In *Hitos Históricos de los Regadíos Españoles*, ed. A. Gil Olcina and A. Morales Gil, 263–307. Madrid: Ministerio de Agricultura, Pesca y Alimentación.

Lopez-Gunn, E. 2009. "*Agua para Todos*: A New Regionalist Hydraulic Paradigm in Spain." *Walter Alternatives* 2 (3): 370–394.

Lorenzo Pardo, M. 1916. *El Pantano del Ebro. Memoria*. Zaragoza: Archivo Confederación Hidrográfica del Ebro.

Lorenzo Pardo, M. 1930. *La Confederación del Ebro*. Madrid: Compañía Ibero-Americana de Publicaciones.

Lorenzo Pardo, M. 1931. *La Conquista del Ebro*. Zaragoza: Editorial Heraldo de Aragón.

Lorenzo Pardo, M. [1933] 1988. *Las Directrices de una Nueva Política Hidráulica y Los Riegos de Levante*. Murcia: Consejería de Política Territorial y Obras Públicas de la Comunidad Autónoma de la Región de Murcia.

Lorenzo Pardo, M. [1933] 1999. *Plan Nacional de Obras Hidráulicas*. Facsimile reissue of 1933 *Hydraulic Plan*. Madrid: Ministerio de Obras Publicas, Transportes y Medio Ambiente.

Machado, A. 1912. *Campos de Castilla. La Lectura*, no. 110 (February 1910). Madrid: Renacimiento. http://www.abelmartin.com/guia/antol/cam_2.html (accessed July 20, 2011).

Macías, A. M., and M. Ojeda. 1989. "Acerca de la Revolución Burguesa y su Reforma Agraria: la Desamortización del Agua." *Anuario de Estudios Atlánticos* 35: 217–261.

Macías Picavea, R. 1895. *Geografía Elemental. Compendio Didáctico y Razonado*. Valladolid: Establecimiento tipográfico de H. de J. Pastor.

Macías Picavea, R. 1896. *La Tierra de Campos*. Madrid: Librería de Victoriano Suárez.

Macías Picavea, R. [1899] 1977. *El Problema Nacional*. Madrid: Instituto de Estudios de Administración Local.

Macnaghten, P., and J. Urry. 1998. *Contested Natures*. London: Sage.

Mainer, J. C. 1972. *Literatura y Pequeña-Burguesía en España Notas 1890–1950*. Madrid: Editorial Cuadernos para el Diálogo.

Mairal Buil, G. 2005. "Los Conflictos del Agua en España." *Nómadas* 22: 126–139.

Mallada, L. 1882. "Causas de la Pobreza de Nuestro Suelo." *Boletín de la Sociedad Geográfica de Madrid* 7: 89–109.

Mallada, L. 1890. *Los Males de la Patria y la Futura Revolución Española*. Madrid: Fundación Banco Exterior.

Maluquer de Motes, J. 1983. "La Despatrimonialización del Agua: Movilización de un Recurso Natural Fundamental." *Revista de Historia Económica* 1 (2): 76–96.

Manteiga, L., and C. Olmeda. 1992. "La Regulación del Caudal Ecológico." *Quercus* 78: 44–46.

MAPA. 1990. *Historia y Evolución de la Colonización Agraria en España—Vol. II—Política Administrativa y Económica de la Colonización Agraria—Análisis Institucional y Financiero (1936–1977)*. Madrid: Ministerio para las Administraciones Públicas.

March Corbella, H. 2010. Urban Water Management and Market Environmentalism: A Historical Perspective for Barcelona and Madrid. Doctoral dissertation, Institut de Ciència i Tecnologia Ambientals—Departament de Geografia, Facultat de Filosofia i Lletres, Autonomous University of Barcelona, Barcelona.

March Corbella, H., and D. Saurí. 2008. Crisis-ridden Water Governance: The Drought of 2008 in Metropolitan Barcelona. Paper presented at RGS-IBG Annual Conference, August 27–29, Manchester University, Manchester, UK.

March, H., D. Saurí, and A. M. Rico-Amorós. 2014. "The End of Scarcity? Water Desalination as the New Cornucopia for Mediterranean Spain." Forthcoming in *Journal of Hydrology*. doi:10.1016/j.hydrol.2014.04.23.

Marcuello, J. R. 1990. *Manuel Lorenzo Pardo*. Madrid: Colegio de Ingenieros de Caminos, Canales y Puertos/iberCaja.

Martín Barajas, S. 2010. "Reducción de Recursos Hídricos en España." *Ecologista*, no. 65: 60–62.

Martínez Alier, J. 1968. *La Estabilidad del Latifundismo*. Paris: Ruedo Ibérico.

Martínez Gil, F. J. 1997. *La Nueva Cultura del Agua en España*. Bilbao: Bakeaz.

Martínez Gil, F. J. 1999. "Nudos Gordianos de las Políticas del Agua en España." In *El Agua a Debate desde la Universidad: Hacia una Nueva Cultura del Agua*, ed. P. Arrojo Agudo and F. J. Martínez Gil, 103–143. Zaragoza: Institución "Fernando el Católico," Excma. Diputación de Zaragoza.

Martín Gaite, C. 1983. *El Conde de Guadalhorce, su Época y su Labor*. Madrid: Colegio de Ingenieros de Caminos, Canales y Puertos-Ediciones Turner.

Martín Martín, V. 2003. "Análisis Económico y Economía Aplicada en el Pensamiento Económico Español de los Siglos XVIII y XIX: A Propósito del Regeneracionismo." In *Estudios de Historia y de Pensamiento Económico: Homenaje al Profesor Francisco Bustelo García del Real*, ed. J. Hernández Andreu, 241–268. Madrid: Editorial Complutense S.A.

Martín Mendiluce, J. M. 1996. "Los Embalses en España: Su Necesidad y Transcendencia Económica." *Revista de Obras Públicas*, no. 3354: 7–24.

Martín Mendiluce, J. M., and C. Torres Padilla. 1982. "Las Presas en España." *Revista de Obras Públicas* 129 (3202): 219–221.

Martín Retortillo, S. 1963. *La Ley de Aguas de 1866. Antecedentes y Elaboración.* Madrid: Ediciones Centro de Estudios Hidrográficos.

Marx, K. 1971. *Capital, Volume 1.* New York: Penguin.

Marx, K., and F. Engels. [1848] 2012. *The Communist Manifesto.* London: Verso.

Masjuan, E., H. March, E. Domene, and D. Saurí. 2008. "Conflicts and Struggles over Urban Water Cycles: The Case of Barcelona 1880–2004." *Tijdschrift voor Economische en Sociale Geografie* 99 (4): 426–439.

Mateu Bellés, J. F. 1994. "Planificación Hidráulica de las Divisiones Hidrológicas 1865–1899." Valencia: Department of Geography, University of Valencia.

Mateu Bellés, J. F. 1995. "Planificación Hidráulica de las Divisiones Hidrológicas." In *Planificación Hidráulica en España*, ed. A. Gil Olcina and A. Morales Gil, 69–106. Murcia: Fundación Caja del Mediterráneo.

Mateu Bellés, J. F. 2004. "Los Aforos de los Ríos Peninsulares durante la Primera Etapa de las Divisiones Hidrológicas (1865–1876)." In *Historia, Clima y Paisaje: Estudios Geográficos en Memoria del Profesor Antonio López Gómez*, ed. V. M. Rosselló, 363–382. Valencia: Universidad de Valencia.

Mateu González, J. J. 2002. "Política Hidráulica e Intervención Estatal en España (1880–1936): Una Visión Interdisciplinar." *Estudios Agrosociales y Pesqueros* 197: 35–61.

McCool, D. 1994. *Command of the Waters: Iron Triangles, Federal Water Development and Indian Water.* Tucson: University of Arizona Press.

Meerganz von Medeazza, G. 2005. "'Direct' and Socially-Induced Environmental Impacts of Desalination." *Desalination* 185 (1–3): 57–70.

Meerganz von Medeazza, G. 2008. *Escasez de Agua Dulce y Desalinización.* Bilbao: Bakeaz.

Melgarejo Moreno, J. 1995. *La Intervención del Estado en la Cuenca del Segura, 1926–1986.* Valencia: Instituto de Cultura "Juan Gil-Albert."

Melgarejo Moreno, J. 2000. "De la Política Hidráulica a la Planificación Hidrológica. Un Siglo de Intervención del Estado." In *El Agua en la Historia de España*, ed. C. Barcielo López and J. Melgarejo Moreno, 273–321. Alicante: Publicaciones de la Universidad de Alicante.

Mendez, R. 2005. "Barreda: "Estamos en una larga guerra del agua que Castilla-La Mancha va a ganar." *El País*, July 5. http://elpais.com/diario/2005/07/05/espana/1120514417_850215.html (accessed July 24, 2011).

Mendoza Gimeno, J. L. 1961. "Los Ingenieros Hidráulicos en España." *Revista de Obras Públicas* (June): 364–367.

Menéndez Prieto, M. n.d. *Política del Agua en España: Principales Actuaciones dese 1996*. Madrid: Ministerio de Agricultura, Alimentación y Medio Ambiente.

Miller, B. A, and R. B. Reidinger 1998. *Comprehensive River Basin Development: The Tennessee Valley Authority*. Washington, DC: World Bank Publications.

Ministerio de Agricultura. 1980. *Anuario de Estadística Agraria*. Madrid: Ministerio de Agricultura.

Ministerio de Agricultura, Alimentación y Medio Ambiente. 2011. "Actuaciones y Proyectos en el Programa A.G.U.A." http://www.magrama.gob.es/es/agua/planes-y-estrategias/informes-de-viabilidad-de-obras-hidraulicas/actuaciones-y-proyectos/index_agua.aspx (accessed July 25, 2011).

Ministerio de Agricultura, Pesca y Alimentación. 2008. *Plan Nacional de Regadíos—Horizonte 2008*. Madrid: Ministerio de Agricultura, Pesca y Alimentación.

Ministerio de Medio Ambiente. 1998. *Libro Blanco del Agua en España*. Madrid: Ministerio de Medio Ambiente.

Ministerio de Medio Ambiente. 2000a. *Plan Hidrológico Nacional, Análisis de Antecedentes y Transferencias Planteadas*. Madrid: Ministerio de Medio Ambiente.

Ministerio de Medio Ambiente. 2000b. *Plan Hidrológico Nacional*. 5 vols. Madrid: Ministerio de Medio Ambiente.

Ministerio de Obras Públicas. 1967. *Anteproyecto General de Aprovechamiento Conjunto de Recursos Hidráulicos del Centro y Sureste de España—Compleja Tajo-Segura*. 2 vols. Madrid: Ministerio de Obras Públicas, Dirección General de Obras Hidráulicas.

Ministerio de Obras Públicas y Transportes. 1993a. *El Plan Hidrológico Nacional—Memoria*. Madrid: Ministerio de Obras Públicas y Transportes, Dirección General de Obras Hidráulicas.

Ministerio de Obras Públicas y Transportes. 1993b. *El Plan Hidrológico Nacional—Memoria, Exposición de Motivos*. Madrid: Ministerio de Obras Públicas y Transportes, Dirección General de Obras Hidráulicas.

Ministerio de Obras Públicas y Urbanismo. 1990. Plan Hidrológico—Sintesis de la Documentación Básica. Madrid: Dirección General de Obras Hidraulicas.

Miranda Encarnación, J. A. 2003. "El Fracaso de la Industrialización Autárquica." In *Autárquica y Mercado Negro. El Fracaso Económico del Primer Franquismo, 1939–1959*, ed. C. Barciela López, 95–121. Barcelona: Crítica.

Mitchell, T. 2002. *Rule of Experts: Egypt, Techno-Politics, Modernity*. Oakland: University of California Press.

Molinero, C., M. Sala, and J. Sobrequés, eds. 2003. *Una Inmensa Prisión: Los Campos de Concentración y las Prisiones durante la Guerra Civil y el Franquismo*. Barcelona: Ed. Crítica.

Molle, F. 2009. "River-Basin Planning and Management: The Social Life of a Concept." *Geoforum* 40 (3): 484–494.

Molle, F., P. P. Mollinga, and P. Wester. 2009. "Hydraulic Bureaucracies and the Hydraulic Mission: Flows of Water, Flows of Power." *Water Alternatives* 2 (3): 328–349.

Morales Amores, A. 1899a. "Intervención del Estado en la Construcción de Pantanos y Canales de Riego." *Revista de Obras Públicas* 46 (1231): 164–166.

Morales Amores, A. 1899b. "Plan de Pantanos y Canales de Riego—Ahora ó Nunca." *Revista de Obras Públicas* 46 (1232): 175–176.

Morales Amores, A. 1899c. "Plan de Pantanos y Canales de Riego—Inconvenientes de Retardar su Formación." *Revista de Obras Públicas* 46 (1233): 193–194.

Moreno Gómez, F. 2008. *1936: el Genocidio Franquista en Córdoba*. Barcelona: Editorial Crítica.

Moreu, J. L. 1999. "El Marco Jurídico de la Política Hidráulica." In *El Agua a Debate desde la Universidad: Por una Nueva Cultura del Agua*, ed. P. Arrojo and J. Martínez Gil, 783–815. Zaragoza: Fundación Fernando el Católico, CSIC.

Morton, T. 2007. *Ecology without Nature*. Cambridge, MA: Harvard University Press.

Murphy, B. 1999. "European Integration and Liberalization: Political Change and Economic Policy Continuity in Spain." *Mediterranean Politics* 4 (1): 53–78.

Nadal Reimat, E. 1981. "El Regadío Durante la Restauración." *Agricultura y Sociedad* 19: 129–163.

Nárdiz Ortiz, C. 2003. "1975–2003. La Transición Tardía de la Revista a la Etapa Actual." *Revista de Obras Públicas*, no. 3434: 103–115.

Natter, W., and W. Zierhofer. 2002. "Political Ecology, Territoriality, and Scale." *GeoJournal* 58 (4): 225–232.

Niño, A. 2003. "50 Años de Relaciones entre España y Estados Unidos." *Cuadernos de Historia Contemporánea* 25: 9–33.

NN. 1899. "Instrucciones para la Servicio Hidrológico en las Provincias." *Revista de Obras Públicas* 46 (1265): 485–486.

NN. 1902a. "Ministerio de Agricultura, Industria, Comercio y Obras Públicas." *Revista de Obras Públicas* 50 (1388): 386–388.

NN. 1902b. "Ministerio de Agricultura, Industria, Comercio y Obras Públicas—Plan de Obras Públicas (continuación)." *Revista de Obras Públicas* 50 (1389): 404–406.

NN. 1902c. "Ministerio de Agricultura, Industria, Comercio y Obras Públicas—Plan de Obras Públicas (conclusión)." *Revista de Obras Públicas* 50 (1390): 423–425.

NN. 1904. "El Servicio Hidráulico." *Revista de Obras Públicas* 52 (1392): 7–10.

NN. 1909. *Plan de Obras Hidráulicas Realizable en un Plazo de Ocho Años.* Madrid: Ministerio de Fomento. Dirección de Obras Públicas.

NN. 1911. "Proyectos de Ley de Obras Publicas." *Revista de Obras Públicas* 59 (1850): 121–142.

NN. 1930. "Editorial." *Montes e Industrias* 1: 29–30.

NN. 1931. "Editorial." *Revista de la Confederación Hidrográfica del Ebro* 5 (47): 1.

NN. 1953. *The Concordat between Spain and the Holy See (27 August 1953).* Madrid: Diplomatic Information Office.

NN. 2005. "La Tragedia de Ribadelago (Zamora)." *La Riada*, no. 10: 7.

NN. 2011. "Marcha azul a Bruselas En defensa de una nueva Cultura del Agua, es decir, del desarrollo sostenible." http://www.pasapues.es/naturalezadearagon/agua/azul/index.php (accessed December 21, 2011).

Norgaard, R. 1994. *Development Betrayed: The End of Progress and a Coevolutionary Revisioning of the Future.* New York: Routledge.

Norman, E. S., K. Bakker, and C. Cook. 2012. "Introduction to the Themed Section: Water Governance and the Politics of Scale." *Water Alternatives* 5 (1): 52–61.

Núñez, G. 1995. "Empresas de Producción y Distribución de Electricidad en España (1878–1953)." *Revista de Historia Industrial*, no. 7: 39–79.

Núñez, G. 2003. "Las Empresas Eléctricas: Crisis de Crecimiento en un Contexto de Crisis Política." In *Los Empresarios de Franco. Política y Economía en España, 1936–1957*, ed. G. Sánchez Recio and T. Fernández, 121–144. Barcelona: Crítica.

Opie, J. 1998. *Nature's Nation: An Environmental History of the United States.* Forth Worth, TX: Harcourt Brace College Publishers.

Ortega, N. 1975. *Política Agraria y Dominación del Espacio—Origines, Caracterización y Resultados de la Política de Colonización planteada en la España posterior a la Guerra Civil.* Madrid: Editorial Ayuso.

Ortega Cantero, N. 1992. "El Plan Nacional de Obras Hidráulicas." In *Hitos Históricos de los Regadíos Españoles*, ed. A. Gil Olcina and A. Morales Gil, 335–364. Madrid: Ministerio de Agricultura, Pesca y Alimentación.

Ortega Cantero, N. 1995. "El Plan General de Canales de Riego y Pantanos de 1902." In *Planificación Hidráulica en España*, ed. A. Gil Olcina and A. Morales Gil, 107–136. Murcia: Fundación Caja del Mediterráneo.

Ortega y Gasset, J. [1921] 2007. *España Invertebrada*. Alianza Editorial: Madrid.

Ortega y Gasset, J. 1983. "Sobre la Vieja Política, *El Sol*, 27 noviembre 1923." In *Obras Completas*, ed. J. Ortega y Gasset, 26–31. Madrid: Editorial Alianza. Ortí, A. 1976. "Infortunio de Costa y Ambigüedad del Costismo: Una Reedición Acrítica de 'Política Hidráulica.'" *Agricultura y Sociedad*, no. 1: 179–190.

Ortí, A. 1984. "Política Hidráulica y Cuestión Social: Orígenes, Etapas y Significados del Regeneracionismo Hidráulico de Joaquín Costa." *Revista Agricultura y Sociedad* 32: 11–107.

Ortí, A. 1994. "Política Hidráulica y Emancipación Campesina en el Discurso Político del Populismo Rural Español entre las dos Repúblicas contemporáneas." In *Regadíos y Estructuras de Poder*, ed. J. Romero and C. Giménez, 241–267. Alicante: Instituto de Cultura "Juan Gil-Albert," Diputación de Alicante.

Pachauri, R. K., and A. Reisinger, eds. 2007. *Contribution of Working Groups I, II and III to the Fourth Assessment Report of the Intergovernmental Panel on Climate Change*. Geneva: Intergovernmental Panel on Climate Change.

Padovan, D. 2000. "The Concept of Social Metabolism in Classical Sociology." *Theomai*, no. 2. http://revista-theomai.unq.edu.ar/numero2/artpadovan2.htm (accessed April 7, 2013).

Page, B., and M. Kaika. 2003. "The EU Water Framework Directive: Part 2. Policy Innovation and the Shifting Choreography of Governance." *European Environment* 13 (6): 328–343.

Palacio, M. 2005. "Early Spanish Television and the Paradoxes of a Dictator General." *Historical Journal of Film, Radio and Television* 25 (4): 599–617.

Palancar Penella, M. 1960. "Comentarios acerca de las Comisarias de Aguas." *Revista de Obras Públicas* (July): 531–532.

Palancar Penella, M. 1976. "Sobre la Necesidad de una Nuevo Planteamiento en el Aprovechamiento Integral de los Recursos Hidráulicos Nacionales." *Revista de Obras Públicas* 123 (3135): 613–616.

Palancar Penella, M. 1979. "¿Es Necesario Seguir Construyendo Presas?" *Revista de Obras Públicas* 126 (3176): 1021–1026.

Palau, A. 2003. *Régimen Ambiental de Caudales: Estado del Arte*. Madrid: Universidad Politécnica de Madrid.

Payne, S. G. 1987. *The Franco Regime: 1936–1975*. Madison: University of Wisconsin Press.

Pearson, D. 1952. "Lobbyists Making Foreign Policy." *St. Petersburg Times*, June 14, 6.

References

Peña Boeuf, A. 1946. "Desarrollo de las Obras Públicas en España—Conferencias Pronunciadas en la Escuela de Ingenieros de Caminos, Canales y Puertos, durante el Mes de Marzo de 1946, por el Profesor Excmo. Sr. D. Alfonso Peña Boeuf, Académico de la Real de Ciencias Exactas, Físicas y Naturales." *Revista de Obras Públicas* (July): 357–371.

Peña Boeuf, A. 1955. "Los Riegos en España." *Revista de Obras Públicas*, no. 2888: 613–616.

Pérez, J. 1999. *Historia de España*. Barcelona: Crítica.

Pérez Crespo, A. 2009. *Los Origines y Puesta en Marcha Del Trasvase Tajo-Segura—Una Crónica Personal*. Murcia: Fundación Instituto Euromediterráneo del Agua.

Pérez de la Dehesa, R. 1966. *El Pensamiento de Costa y su Influencia en el 98*. Madrid: Editorial Sociedad de Estudios y Publicaciones.

Pérez Díaz, V., and J. Mezo. 1999. "Política del Agua en España: Argumentos, Conflictos y Estilos de Deliberación." In *El Agua a Debate desde la Universidad—Hacia una Nueva Cultura del Agua*, ed. P. Arrojo Agudo and F. J. Martínez Gil, 625–648. Zaragoza: Institución "Fernando el Católico," Excma. Diputación de Zaragoza.

Pérez Díaz, V., J. Mezo, and B. Álvarez-Miranda. 1996. *Política y Economía del Agua en España*. Madrid: III Premio Círculo de Empresarios.

Pérez Picazo, M. T. 1999. "Gestión del Agua y Conflictividad en el Sureste de España, Siglos XIX y XX." In *El Agua a Debate desde la Universidad—Hacia una Nueva Cultural del Agua*, ed. P. Arrojo Agudo and F. J. Martínez Gil, 649–667. Zaragoza: Institución "Fernando el Católico," Excma. Diputación de Zaragoza.

Pérez Picazo, M. T., and G. Lemeunier. 2000. "Formation et Mise en Crise du Modèle de Gestion Hydraulique Espagnol de 1780 à 2000." *Économies et Societés* 37: 71–98.

Pons Múria, M. 2003. *Water: The Source of Life*. Video documentary. Produced by M. Pons Múria, M. Carreras, and A. Múria. Zaragoza: Platform for the Defence of the River Ebro.

Preston, P. 1990. *The Politics of Revenge: Fascism and the Military in 20th Century Spain*. London: Unwin Hyman.

Preston, P. 1995. *Franco: A Biography*. London: Fontana Press.

Preston, P., and F. Lannon, eds. 1990. *Elites and Power in Twentieth-Century Spain*. Oxford: Clarendon Press.

Prieto, I. 1933. "Politica Hidráulica—Plan Nacional." In *Plan Nacional de Obras Hidráulicas*, ed. M. Lorenzo Pardo, 1–3. Madrid: Centro de Estudios Hidrográficas.

Programa A.G.U.A. 2010. *Dossier Bibliográfico de Publicaciones sobre Política y Gestión del Agua*. Programa A.G.U.A. http://www.ecoestrategia.com/articulos/hemeroteca/politicaagua.pdf. (accessed August 21, 2013).

Puente Diaz, G. 1949. "La Energía Eléctrica en España." *Información Comercial Española* (Numero extraordinario dedicada a la Electricidad Española): 55–59.

Puig, N. 2003. "La Ayuda Económica Norteamericana y los Empresarios Españoles". *Cuadernos de Historia Contemporánea* 25: 109–129.

Puig, N., and A. Alvaro. 2002. "Estados Unidos y la Modernización de los Empresas Españoles, 1950–1975: Un Estudio Preliminar." *Historia del Presente*, no. 1: 9–29.

Puig y Valls, R. 1898. "La Patria y el Árbol. Síntesis de un Proyecto y de su Inmediata Ejecución." *La Vanguardia*, September 21.

Puig y Valls, R. 1909. *Crónica de la Fiesta del Árbol en España*. Barcelona: Talleres Gráficos de José Casamajó.

Ramírez Ruiz, R. 2008. *Caciquismo y Endogamia. Un Análisis del Poder Local en el España de la Restauración*. Madrid: Editorial Dykinson, S.L.

Ramón y Cajal, S. 1899. *Reglas y Consejos sobre Investigación Biológica*. Madrid: Imprenta de Fortanet.

Ramos Gorostiza, J. L. 2001. "La Formulación de la Política Hidrológica en el Siglo XX: Ideas e Intereses: "Actores" y Proceso Político." In *Documentos de Trabajo de la Facultad de Ciencias Económicas y Empresariales*, 28 pp. Madrid: Universidad Complutense de Madrid, Facultad de Ciencias Económicas y Empresariales.

Reguera Rodríguez, A. T. 1991. "Fascismo y Geopolítica en España." *GeoCrítica*, no. 94: 11–63.

Reher, D. S. 2003. "Perfiles Demográficos de España, 1940–1960." In *Autarquía y Mercado Negro—El Fracaso Económico del Primer Franquismo, 1939–1959*, ed. C. Barciela López, 1–26. Barcelona: Crítica.

Richards, M. 1999. *Un Tiempo de Silencio. La Guerra Civil y la Cultura de la Represión en la España de Franco*. Barcelona: Crítica.

Rico Amorós, A. M. 2010. "Plan Hidrológico Nacional y Programa A.G.U.A.: Repercusión en las Regiones de Murcia y Valencia." *Investigaciones Geográficas*, no. 51: 235–267.

Rodríguez, S. 1999. *El NO-DO: Catecismo Social de una Época*. Madrid: Editorial Complutense.

Rodríguez Ferrero, N. 2001. *Los Regadíos de Iniciativa Pública en la Cuenca del Guadalquivir: un Análisis Económico*. Granada: Editorial Universidad de Granada.

ROP. 1923 "Editorial." *Revista de Obras Públicas* 71 (2390): 86.

ROP. 1930. "Editorial." *Revista de Obras Públicas* 78 (2543): 68.

ROP. 1931. "Editorial." *Revista de Obras Públicas* 79: 166.

ROP. 1936. "Fijando Posiciones." *Revista de Obras Públicas* 84 (2699): 1.

ROP. 1940. "Número Extraordinario, dedicado a la Cruzada Española. 1936–1939." *Revista de Obras Públicas* (Special Issue).

ROP. 1963. "Crónica. Inauguración del Centro de Estudios Hidrográficos por S. E. el Jefe del Estado." *Revista de Obras Publicas* (August): 553.

ROP. 2003a. "Editoriales Políticos: Las Crisis de 1923 a 1939." *Revista de Obras Públicas*, no. 3434: 52–53.

ROP. 2003b. "La Revista y las Ideas de Costa." *Revista de Obras Públicas*, no. 3434: 34–36.

Rosendorf, N. M. 2006. "Be El Caudillo's Guest: The Franco Regime's Quest for Rehabilitation and Dollars after World War II via the Promotion of U.S. Tourism in Spain." *Diplomatic History* 30 (3): 376–407.

Rothenberg, D., and M. Ulvaeus. 2001. *Writing on Water*. Cambridge, MA: MIT Press.

Ruiz, J. M. 1993. "La Situación de los Recursos Hídricos en España, 1992." In *La Situación en el Mundo—1993*, ed. L. R. Brown, 385–444. Madrid: Ediciones Apóstrofe.

Sabio Alcutén, A. 1994. "Herencia de Preguerra, Fachada de Postguerra: Regadío y Obras Públicas en Huesca, 1938–1960." In *Agua y Progreso Social—Siete Estudios sobre el Regadío en Huesca, Siglos XII–XX*, ed. C. Laliena Corbera, 215–250. Huesca: Diputación de Huesca, Instituto de Estudios Altoaragoneses.

Sáenz García, C. 1967. "El Agua y las Demarcaciones Geográficas." *Revista de Obras Públicas* (March): 181–193.

Sáenz Ridruejo, F. 1999. "Los Ingenieros de Caminos—Doscientos Años de Historia." *Revista de Obras Públicas*, no. 3388: 8–15.

Sáenz Ridruejo, F. 2000. Los ingenieros de Caminos de la Generación del 98. *Los Cuadernos del Cauce* 14: 3–15.

Sáenz Ridruejo, F. 2001. "La Ingeniería del Agua en el Siglo XIX." *Revista de Obras Públicas*, no. 3414: 45–49.

Sáenz Ridruejo, F. 2003. "Un Siglo y Medio de La Revista de Obras Públicas." *Revista de Obras Públicas*, no. 3434: 7–15.

Sahuquillo, A. 1993. "Los Objetivos de los Planos Hidrológicos. Las Aguas Subterráneas, la Contaminación y el Medio Ambiente." *Revista de Obras Públicas* 140 (3318): 17–22.

Sánchez de Toca, J. 1911. *Reconstitución de España en Vida de Economía Política Actual*. Madrid: Jaime Ratés Martín Impresor.

Sánchez Illán, J. C. 1898. "El Ascenso Político de la Elite Periodística: Rafael Gasset, Primer Ministro de Agricultura, Industria, Comercio y Obras Públicas." *Studia Historica—Historia Contemporánea* 16: 221–245.

Sánchez Pérez, F. 1981. "Acceso al Profesorado en la Geografía Española (1940–1979)." *GeoCrítica*, no. 32: 5–51.

Sánchez Recio, G. 2003. "El Franquismo come Red de Intereses." In *Los Empresarios de Franco. Política y Economía en España, 1936–1957*, ed. G. Sánchez Recio and J. Tascón Fernánde, 13–22. Barcelona: Crítica.

Sánchez Rey, A. 2003. "El Papel de la Revista de Obras Públicas en la Prensa Española: Un Medio de Información Testigo de Tres Siglos." *Revista de Obras Públicas* (3434): 17–38.

Sánchez-Albornoz, N. 2004. "Saña y Negocio en le Trabajo Forzado." In *El Canal de los Presos (1940–1962). Trabajos Forzados: De la Represión Política a la Explotación Económica*, ed. G. Acosta Bono, J. L. Gutiérrez Molina, L. Martínez Macías, and A. del Río Sánchez, xxi–xxv. Barcelona: Crítica.

Sancho de Ybarra, G. 1977. "Gestión Autónoma del Uso del Agua en España." *Revista de Obras Públicas* 124 (3149): 687–700.

Sancho Portero, I. 2008. *El Trasvase Tajo-Segura: Debate, Impacto y Propuestas*. Murcia: Universidad de Murcia.

Saraiva, T., and M. N. Wise. 2010." Autarky/Autarchy: Genetics, Food Production, and the Building of Fascism." *Historical Studies in the Natural Sciences* 40 (4): 419–428.

Sardá, J. 1970. "El Banco de España (1931–1962)." In *El Banco de España. Una Historia Económica*, ed. Banco de España, 419–479. Madrid: Banco de España.

Saurí, D., and L. del Moral. 2001. "Recent Developments in Spanish Water Policy. Alternatives and Conflicts at the End of the Hydraulic Age." *Geoforum* 32: 351–362.

Schlosberg, D. 2007. *Defining Environmental Justice: Theories, Movements, and Nature*. Oxford: Oxford University Press.

Schmidt, G. 2010. "España Desala el Oro de sus Costas." http://www.tetrapak.com/es/documents/2006_Juan_Monzon.pdf (accessed August 23, 2013).

Schnaiberg, A. 1980. *The Environment: From Surplus to Scarcity*. New York: Oxford University Press.

Selby, J. 2003. *Water, Power and Politics in the Middle East: The Other Israeli-Palestinian Conflict*. London: I. B. Taurus.

Senador Gómez, J. 1915. *Castilla en Escombros: las Leyes, las Tierras, el Trigo y el Hambre; Los Derechos del Hombre y los del Hambre*. Valladolid: Casa Editorial y Libreria de la Viuda De Montero.

References

Serrada, R. 2000. *Apuntes de Repoblaciones Forestales.* Madrid: FUCOVASA.

Serrano, C. 1984. *Final del Imperio: España 1895–1898.* Madrid: Siglo XXI.

Shapiro, J. 2001. *Mao's War against Nature: Politics and the Environment in Revolutionary China.* Cambridge: Cambridge University Press.

Simpson, J. 1995. *Spanish Agriculture: The Long Siesta, 1765–1965.* Cambridge: Cambridge University Press.

Smith, J. 1994. *The Spanish-American War: Conflict in the Caribbean and the Pacific 1895–1902.* London: Longman.

Smith, N. 1984. *Uneven Development.* Oxford: Blackwell.

Smith, N. 1987. "Rehabilitating a Renegade? The Geography and Politics of Karl August Wittfogel." *Dialectical Anthropology* 12 (1): 126–136.

Smith, N. 1996. "The Production of Nature." In *FutureNatural: Nature/Science/Culture,* ed. G. Robertson, L. Mash, L. Tickner, J. Bird, B. Curtus, and T. Putnam, 35–54. London: Routledge.

Smith, N. 2006. "Nature as an Accumulation Strategy." In *Coming to Terms with Nature. Socialist Register 2007,* ed. L. Pantitch and C. Leys, 16–36. London: The Merlin Press.

Smith, T. C. 1969. "The Drainage Basin as an Historical Unity for Human Activity." In *Introduction to Geographical Hydrology,* ed. R. J. Chorley, 20–29. London: Methuen.

Sneddon, C. 2003. "Reconfiguring Scale and Power: The Khong-Chi-Mun Project in Northeast Thailand." *Environment and Planning A* 35 (12): 2229–2250.

Solé Villalonga, G. 1967. *La Reforma fiscal de Villaverde, 1899–1900.* Madrid: Editorial de Derecho Financiero.

Soler, J. G. 2002. "Los Esclavos del Franquismo." *El Siglo.* http://www.foroporlamemoria.info/documentos/esclavos_franquismo.htm (accessed April 1, 2011).

Songel González, J. M. 2003. "La Revista de Obras Públicas entre 1939 y 1959: Desde la Autarquía al Plan de Estabilización." *Revista de Obras Públicas,* no. 3434: 83–98.

Suárez Fernández, L. 1984. *Francisco Franco y Su Tiempo.* Madrid: Fundación Nacional Francisco Franco.

Sudriá, C. 1997. "La Restricción Energética al Desarrollo Económico de España." *Papeles de Economía Española,* no. 73: 165–188.

Sudriá i Triay, C. 1990. "La Industria Eléctrica y el Desarrollo Económico de España." In *Electricidad y Desarrollo Económico: Perspectiva Histórica de un Siglo,* ed. J. L. Garcia Delgado, 147–184. Oviedo: Hidroeléctrica del Cantábrico, S.A.

Swyngedouw, E. 1996. "The City as a Hybrid: On Nature, Society and Cyborg Urbanisation." *Capitalism, Nature, Socialism* 7 (1): 65–80.

Swyngedouw, E. 1997. "Neither Global nor Local: "Glocalization" and the Politics of Scale." In *Spaces of Globalization: Reasserting the Power of the Local*, ed. K. Cox, 137–166. New York: Guilford.

Swyngedouw, E. 2003. "Scaled Geographies. Nature, Place, and the Politics of Scale." In *Scale and Geographic Inquiry: Nature, Society and Method*, ed. R. McMaster and E. Sheppard, 129–153. Oxford and Cambridge, MA: Blackwell.

Swyngedouw, E. 2004a. *Social Power and the Urbanisation of Water: Flows of Power.* Oxford: Oxford University Press.

Swyngedouw, E. 2004b. "Globalisation or "Glocalisation"? Networks, Territories and Rescaling." *Cambridge Review of International Affairs* 17 (1): 25–48.

Swyngedouw, E. 2006a. "Metabolic Urbanization: The Making of Cyborg Cities." In *In the Nature of Cities: Urban Political Ecology and the Politics of Urban Metabolism*, ed. N. Heynen, M. Kaika, and E. Swyngedouw, 21–40. London: Routledge.

Swyngedouw, E. 2006b. "Circulations and Metabolisms: (Hybrid) Natures AND (Cyborg) Cities." *Science as Culture* 15 (2): 105–122.

Swyngedouw, E. 2007. "Impossible/Undesirable Sustainability and the Post-Political Condition." In *The Sustainable Development Paradox*, ed. R. Krueger and D. Gibbs, 13–40. New York: Guilford Press.

Swyngedouw, E. 2010a. "Apocalypse Forever? Post-Political Populism and the Spectre of Climate Change." *Theory, Culture, Society* 27 (2–3): 213–232.

Swyngedouw, E. 2010b. "Trouble with Nature: Ecology as the New Opium for the People." In *Conceptual Challenges in Planning Theory*, ed. J. Hillier and P. Healey, 299–320. Farnham: Ashgate.

Swyngedouw, E. 2011. "Depoliticized Environments: The End of Nature, Climate Change and the Post-Political Condition." *Royal Institute of Philosophy Supplement* 69: 253–274.

Tabara, D., and A. Ilhan. 2008. "Culture as Trigger for Sustainable Transition in the Water Domain: The Case of the Spanish Water Policy and the Ebro Riber Basin." *Regional Environmental Change* 8 (2): 59–71.

Tamames, R., and X. Casals. 2004. *Miguel Primo de Rivera y Orbaneja.* Barcelona: Ediciones B.

Termis Soto, F. 2005. *Renunciando a Todo: El Régimen Franquista y los Estados Unidos desde 1945 hasta 1963.* Madrid: Biblioteca Nueva.

Thiel, A., D. Sampedro, and C. Schröder. 2011 "Explaining Re-scaling and Differentiation of Water Management on the Iberian Peninsula." Paper presented at the VII

References

Congreso Ibérico sobre Gestión y Planificación del Agua "Ríos Ibéricos +10. Mirando al futuro tras 10 años de DMA," February 16–19, Talavera de la Reina, Spain.

Thorning, J. F. [1943] 1968. *Builders of the Social Order*. Freeport, NY: Books for Libraries Press.

Tierno Galván, E. 1961. *Costa y el Regeneracionismo*. Barcelona: Editorial Barna.

Torán, J. 1971. "Visita al Jefe del Estado del Comité Español de Grandes Presas." *Revista de Obras Públicas* (April): 314–315.

Torán, J., and J. A. Herreras. 1977. "Las Grandes Presas en el Desarrollo de los Recursos Hidráulicos. La Experiencia Española." *Revista de Obras Públicas* (April): 259–266.

Torres Campos, R. 1907. "Nuestros Ríos." *Boletín de la Sociedad Geográfica de Madrid* 37: 7–32.

Torres Martínez, M. 1961. *El Regadío Murciano, Problema Nacional*. 2nd. ed. Murcia: Instituto de Orientación y Asistencia Técnica del Sureste.

Tortella, G. 1981. "La Economía Española 1830–1900." In *Historia de España*. Vol. VIII, ed. M. Tuñón de Lara, 9–167. Madrid: Ed. Labor.

Tranche, R. R., and V. Sánchez-Biosca. 2002. *NO-DO: El Tiempo y la Memoria*. Madrid: Cátedra/Filmoteca Española.

Tuñón de Lara, M. 1971. *Medio Siglo de Cultura Española 1885–1936*. Madrid: Editorial Tecnos.

Tuñón de Lara, M. 1974. *Costa y Unamuno en la Crisis de Fin de Siglo*. Madrid: Editorial Cuadernos para el Diálogo.

Turró, M. 1996. "Capitales Privados—Subvenciones Comunitarias." *Revista de Obras Públicas*, no. 3351: 75–79.

Unión de Pequeños Agricultores. 1993. *Alegaciones de UPA al Plan Hidrológico Nacional*. Madrid: Unión de Pequeños Agricultores.

United Nations Security Council. 1946. *Report of the Subcomittee on the Spanish Question*. New York: UNSC.

Urbistondo, R. 1963. "Centro de Estudios Hidrográficos." *Revista de Obras Públicas* (September): 593–585.

Urbistondo Echeverría, R. 1984. "La Ingeniería Hidráulica: Manuel Lorenzo Pardo." In *La Ingeniería Española en el Siglo XX*, ed. G. Millán Barbany, E. Sánchez-Monge y Parellada, C. Benito Hernández, J. Warleta Carrillo, E. Alastrué Castillo, and R. Urbistondo Echeverría, 177–218. Madrid: Fundación Juan March.

U.S. Department of State. 1953. Department of State Bulletin, U.S. Department of State, Washington, DC. October 5.

Valdés, J. M. 1973. "El Futuro de las Presas en España." *Revista de Obras Públicas* (June): 403–410.

Valera Ortega, J. 2001. *Los Amigos Políticos: Partidos, Elecciones y Caciquismo en la España de la Restauración (1875–1900)*. Madrid: Marcial Pons, Ediciones de Historia, S.L.

Valero de Palma Manglano, J. 1999. "Las Sociedades de Cuenca: Una Puerta Nueva que se Abre a los Usuarios." *Revista de Obras Públicas*, no. 3392: 10–11.

Valladares, S. 1998. "Hacia otra España del Joven Maetzu." *Revista de Antropología Social* 7: 177–213.

Vallarino, E. 1992. "Política Hidráulica." *Revista de Obras Públicas* (March–April): 67–72.

Vallarino Cánovas del Castillo, E., and L. Cuesta Diego. 1999. "Los Aprovechamientos Hidroeléctricos." *Revista de Obras Públicas*, no. 3388: 119–126.

Vallarino Cánovas del Castillo, E., and L. Garrote de Marcos. 2000. "Posibilidades de Aprovechamiento. Límites de la Regulación." *OP: Revista del Colegio de Ingenieros de Caminos, Canales y Puertos* 50: 54–63.

Varela, J. 1997. "Un Profeta Político. Joaquín Costa." *Revista de Historia Económica* 15: 177–184.

Vargas-Amelin, E., and P. Pindado. 2013. "The Challenge of Climate Change in Spain: Water Resources, Agriculture and Land." *Journal of Hydrology*. doi:10.1016/j.jhydrol.2013.11.035.

Velarde Fuertes, J. 1973. *Política Económica de la Dictadura*. Madrid-Barcelona: Giadiana de Publicaciones.

Vera Aparicio, J. A. 2009. *La Gestión Institucional del Agua en España 1978–2008. Una Constitución en Papel Mojado*. Madrid: Ediciones Liteam.

Vera Rebollo, J. F. 1995. "Competencia de Usos y Planificación Fluvial." In *Planificación Hidráulica en España*, ed. A. Gil Olcina and A. Morales Gil, 307–378. Murcia: Fundación Caja del Mediterráneo (CAM).

Vicens Gomez-Tortosa, J. A. 1961. "Resultados Obtenidos hasta la Fecha con la Regulación Hidráulica en España." *Revista de Obras Públicas* (June): 436–441.

Vicens Vives, J. 1940. *España: Geopolítica del Estado y del Imperio*. Barcelona: Editorial Yunque.

Vicens Vives, J. 1941a. "Algunes Caracteres Geopolíticos de la Expansión Mediterránea de España." *Geopolítica* 19 (1): 5–11.

Vicens Vives, J. 1941b. "Spanien und die Geopolitische Neuordnung der Welt." *Zeitschrift für Geopolitik* 18 (5): 256–263.

Vicens Vives, J. 1981. *Tratado General de Geopolítica*. 5th ed. Barcelona: Editorial Vicens Vives.

Vilar, P. 1968. *Catalunya dins l'Espanya Moderna: Recerques sobres els Fonaments Económics de les Estructures Nacionals*. Barcelona: Curial Edicions.

Villanueva Larraya, G. 1987. "Rafael Gasset: "La Política Hidráulica" en la Acción de Gobierno." *Revista de la Facultad de Geografía e Historia* (1): 439–459.

Villanueva Larraya, G. 1990. "Apuntes pare una Biografía política de Rafael Gasset, un Liberal Regeneracionista." *Espacio, Tiempo y Forma, Serie V* 3: 159–171.

Villanueva Larraya, G. 1991. *La "Política Hidráulica" durante la Restauración (1874–1923)*. Madrid: Universidad Nacional de Educación a Distancia.

Viñas, A. 1981. *Los Pactos Secretos de Franco con Estados Unidos. Bases, Ayuda Económica, Recortes de Soberanía*. Barcelona: Ediciones Grijalbo.

Virilio, P. 1986. *Speed and Politis: An Essay on Dromology*. New York: Semiotext(e).

Walker, G. 2011. *Environmental Justice: Concepts, Evidence and Politics*. London: Routledge.

Walters, V. 1993. "El Acuerdo sobre los Bases entre España y Estado Unidos Cuarenta Años Después." *Política Exterior* 36 (7): 160–175.

Warner, J. 2008. "Contested Hydrohegemony: Hydraulic Control and Security in Turkey." *Water Alternatives* 1 (2): 271–288.

White, R. 1995. *The Organic Machine*. New York: Hill and Wang.

White, D. F., and C. Wilbert, eds. 2009. *Technonatures: Environments, Technologies, Spaces, and Places in the Twenty-First Century*. Waterloo, ON: Wilfrid Laurier University Press.

Whitehead, M., R. Jones, and M. Jones. 2007. *The Nature of the State: Excavating the Political Ecologies of the Modern State*. Oxford: Oxford University Press.

Whiteley, J. M., H. Ingram, and R. W. Perry, eds. 2008. *Water, Place, and Equity*. Cambridge, MA: MIT Press.

Wittfogel, K. 1957. *Oriental Despotism. A Comparative Study of Total Power*. New Haven: Yale University Press.

World Health Organization. 2011. *Safe Drinking-water from Desalination. Guidance on Risk Assessment and Risk Management Procedures to Ensure the Safety of Desalinated Drinking-water*. Geneva: WHO.

World Widlife Fund. 1994. *Posición del WWF España Fondo Mundial para la Naturaleza (Adena) sobre Política de Aguas y el Plan Hidrológico Nacional (PHN)*. Madrid: WWF España.

Worster, D. 1985. *Rivers of Empire: Water, Aridity and the Growth of the American West*. New York: Pantheon.

Zapatero, J. L. R. 2006. Discurso de Investidura de José Luis R. Zapatero, 15 de abril 2004. In *Zapatero, en Nombre de Nada—Crónicas y Conversaciones sobre una Deconstrucción*, ed. F. de Haro Izquierdo, 247–270. Madrid: Ediciones Encuentro.

Zimmerer, K. 2000. "Re-scaling Irrigation in Latin America: The Cultural Images and Political Ecology of Water Resources." *Ecumene* 7 (1): 1–35.

Žižek, S. 1989. *The Sublime Object of Ideology*. London: Verso.

Žižek, S. [1992] 2002. *Looking Awry: An Introduction to Jacques Lacan through Popular Culture*. Cambridge, MA: MIT Press.

Index

The letter *f* following a page number denotes a figure, and the letter *t* denotes a table.

Abeima company, 205
Abu Dhabi Investment Authority, 205
Acciona, 205
Actor network theory, hydro-social cycle and, 32–33
AcuaEbro, 214
AcuaMed (*Aguas de las Cuencas Mediterráneas, S.A.*) (Waters of the Mediterranean Basins Ltd.), 196–199, 202–203, 214
AcuaNorte, 214
AcuaSur, 214
Adame, Francisca, 137–139
Agricultural Chamber of Aragón, 74, 236n5
Agriculture
 autarchic control of hydraulic development and, 136–139
 dam construction and food production and, 97–98
 democratization and policies involving, 168–171
 Franco regime's water policies and, 106–108
 land distribution and water politics and, 68–70
 resistance to water transfers in, 177–178
 Spanish Civil War and, 109–111
 Spanish modernization and role of, 44–51
 state involvement hydraulic metabolism and, 76–85
 U.S. aid and development of, 148–149
Albareda, José Luis, 72
Albornoz, Alvaro de, 95
Alfonso XIII (King), 89–90, 94
Almodóvar, Pedro, 167
al Qaeda, Madrid terrorist bombing linked to, 191
Altamira, Rafael, 49
Alzola, Pablo de, 81
American Express Company, 145
American Water Company, 205
Anarchist movement in Spain, early twentieth-century growth of, 69–70
Anfuso, Victor, 147–148
Anti-communism of Franco regime, U.S.-dominated Western Alliance and, 16
Aragón, Autonomous Community of, 179–180
Aridity, hydro-social cycle and sublimity of, 200–202
Artajo, Alberto Martín, 146
Asían Peña, J. L., 121
Australia, hydro-social interventions and national identity in, 31–33
Autarchic development, national hydraulic modernization and geographical integration and, 135–139

Autonomous Communities
  (*Communidades Autónomas*), 179–185
*Avance de un Plan General de Pantanos y
  Canales de Riego* (Proposal for a
  General Plan of Dams and Irrigation
  Canals), 77–85
Aviazione Legionaria (Fascist Italy), 142
Axis powers, Franco's alliance with, 129,
  136, 142–143, 146
Aznar, José María, 182, 186–187, 189,
  191, 195
Azorín. *See* Ruiz, José Martínez (Azorín)

Bacon, Francis, 231n3
Bakker, Karen, 183, 185
Barcelona
  economic "miracle" in, 181
  water supply in, 73
Bardem, Javier, 129
Bardem, Juan Antonio, 129
Barkley, Alben, 144
Baroja, Pío, 54–56, 80
Barreda, José María, 215
Basque Country
  capitalist industrialization in, 40–44
  democratization and autonomy of,
    167, 179
  industrialization in, 44–45
  Itoiz Dam project in, 163
  separatist movement in, 167, 191
Befesa company, 205
Beltrán, Miguel, 88
Ben-Ami, Schlomo, 89
Benjumea Burín, Rafael, 89–91, 112,
  114, 237n20
Bennett, Jean, 28, 200–202
¡Bienvenido Mr Marshall! (Welcome Mr.
  Marshall!) (film), 129–132, 130f,
  242n1
Biopolitics, mercantilization of water
  policy and, 206–207
Blair, Tony, 181
Blanco Rodríguez, Benigno, 183–184

"Blue March" protest, 189
Borrell, Josep, 163, 175
Brandenburg Gate, anti-dam protests at,
  163
Brecht, Bertolt, 223
Brenan, Gerald, 127
British Telecom Pension Scheme, 205

Caciquismo (Spanish political system),
  41–44, 231n2, 233n18
  hydraulic engineering and declining
    power of, 64–66
  Primo de Rivera and, 88
Campo, Julián, 168
Campos de Soria (Machada), 53
Camprubí, Lino, 109
Canada, hydro-social interventions and
  national identity in, 31–33
Canal de Isabel II, 73
Canal del Bajo Guadalquivir, 137
Canal de los Presos (Canal of the
  Prisoners), 137–139, 138f
Canal de Urgel, 73
Canal system, water supply in Spain
  and, 73–74
Cantero, Ortega, 48
Capitalism, historical-geographical
  materialism and, 25–28
Carboneras project, 208
Carr, Raymond, 136
Casado de Otaola, Santos, 46–51
Casares, Francisco, 119
Catalonia
  capitalist industrialization in, 40–44
  democratization and autonomy of,
    167, 179
  regional autonomy *vs.* desalination
    technology in, 215–217
Cavestany de Anduaga, Rafael, 106,
  136
Cenajo dam, 104, 105f
Centre for Metropolitan Planning and
  Research, 181

Index

Centre of Hydrographical Studies
  (*Centro de Estudios Hidrográficos*
  [CEH]), 96–98, 151–152
Centro de Estudios y Experimentación
  de Obras Públicas, 204
Charles V (Holy Roman Emperor and
  King of Spain), 231n3
China. *See* People's Republic of China,
  nation-state formation and
  modernization in
China Investment Corporation, 205
Circulation, hydro-social
  transformation and, 26–28
Citizens' Forum (Murcia), 216–217
Clark, Charles Patrick, 144
Class struggle
  historical-geographical materialism,
    25–28
  in postcolonial Spain, 40–44
Climate change
  desalination and, 220–221
  Spanish water poilitics and, 202–206
Cold War politics
  Franco regime and, 16, 100–101,
    135–139
  U.S.-Spanish relations and, 142–149
Commission of the Geological Map of
  Spain, 49
Commodification of water, neoliberal
  ideology and, 182–185
Common Agricultural Policy, 177
Conde del Guadalhorce. *See* Benjumea
  Burín, Rafael
Condor Legion (Nazi Germany), 142
Confederaciónes hidrográficas, formation
  of, 95
Confederaciones sindicales hidrográficas
  (CSH) (hydrographic
  confederations), 86–88, 91–93
  hydraulic despotism and, 93–94
Congreso de Agriculturos, 47
Coordinadora de Itoiz (Itoiz Coordinating
  Committee), 163, 164f, 165–166

Coordinadora de Organizaciones de
  Defensa Ambiental 1994 (CODA),
  177–178
"Copernican Revolution" in Spanish
  water politics, 91–93
Corps of Engineers (Spain), 57–64
  Franco regime and, 116–119, 120t
  hydrological divisions and, 86–88
  integrated water system proposals and,
    172–176, 174f
  political power and status of, 59–64
  Primo de Rivera dictatorship and,
    89–93
  regeneracionismo and, 42–44
  state involvement hydraulic
    metabolism and, 77–85
Costa Martínez, Joaquin, 65
  desalination technology and legacy of,
    193–194
  Generation of '98 and, 54
  on geography and politics, 46–51, 70
  integrated water system proposals and,
    175–176
  role of state in Spain and, 74–76
  on water scarcity, 51–53
Cronon, William, 8–9
Cuba, Spanish colonization of, 40–44
Cuerpo Especial (Special Corps),
  hydraulic engineers' status as, 59–64
Cultural transformation
  democratization and, 167
  desalination technology and,
    214–217
Cummings, Stanley, Truitt and Cross
  (law firm), 144
Cummings, Homer, 144
Cyborg metaphor
  heterogeneous assemblages and,
    35–37
  hydro-social narrative and, 28–30
  modernization and, 8–9
  politicization of matter and, 29–30
  production of nature and, 20–24

Index

Dam construction
  collapse of dams and, 29–30
  forced labor for, 137–139, 140t–141t
  Franco regime and, 100–101, 107–108, 116–119, 132
  hydro-electrification projects and, 111–116, 113f
  integrated water system plans and, 171–176, 174f
  legacy of Franco's policies in, 157, 158f, 159
  National Hydrological Plan (2000) and, 186–187, 188f
  post-Franco revival of, 163–190
  propaganda films on Franco's inauguration of, 124–127, 126f
  protests against, 163, 164f, 165–166
  in Spain, 11–14, *12–13*
  Spanish hydraulic politics and, 51–53, 60–64, 96–98
  state-sponsored construction, 72–74
  U.S. aid for, 146–149
Dam of Puentes, collapse of, 59
Death of a Cyclist (film), 129
de Cervantes López, Miguel, 86
Deforestation, hydraulic engineering vs., 57–64
De-growth movement, emergence of, 220–221
Del Corral, Enrique, 104
Deleuze, Gilles, 19
De los Rios, Félix, 151
Democratization of Spain
  Franco's opposition to, 148
  hydraulic infrastructure during, 14
  hydro-social politics and, 17
  hydro-structural regime's endurance and, 167–171
  national water planning debate and, 165–166
Depatrimonialization of water, Spanish modernization and, 70–74
De Rivera, Primo, 65

Desalination technology
  climate change and, 202–206
  construction of, 199–200
  engineering, procurement, and construction contractors, 211f
  global capacity and, 208f
  hydro-politics and, 192–221, 197t, 198t, 199f
  international capacity comparisons, 209, 210f
  membrane and thermal capacity and, 209–210, 209f
  membrane-technology contractors, 211–212, 212f
  mercantilization of water policy and, 206–207
  modernization and globalization and, 207–213
  reality and contradictions in, 219–221
  regional and local issues involving, 214–217
  Spanish companies involved in, 209–213, 213f
de Ugalde, Fray Francisco, 231n3
Development, hydro-infrastructure for, 5–6
Diaz-Marta Pinilla, M., 96
Dirección General de Obras Hidráulicas (DGOH), 133–135, 171
División Azul (Blue Division), 142
Draft Proposal for the National Hydrological Plan (1993), 172–178, 174f
  regional river basins and, 179–185
Dragados y Construcciones, 139
Driever, Steven, 45
Drought
  agricultural production and, 77
  climate change and, 202–206
  early hydraulic modernization efforts and, 70
  Franco regime and role of, 103–104, 106, 114

Index

hydro-social cycle and role of, 5, 29
interregional water wars and, 179–185
modernization and role of, 51
public hydraulic projects and, 83–84
Dunn, James, 146

Ebro River Basin Authority
    (*Confederación del Ebro*), 32, 91–93,
    151, 191–192, 216–217, 238n24
Ecological modernization ideology
    climate change and, 203–206
    desalination technology and,
        194–221
    integrated water system proposals and,
        176
    mercantilization of water policy and,
        206–207
    regional autonomy and, 214–217
    sublimity of aridity and, 200–202
Economic Assistance Agreement
    (United States and Spain), 146
Economic conditions
    democratization and, 167–171
    under Franco regime, 135–139
    "Third Way" politics and, 181–182
    U.S. aid to Spain and, 146–149
Education, depoliticization of, 122–123
Egypt, nation-state formation and
    modernization in, 31–33
Elcano, Juan Sebastian, 121
El Chorro Dam, construction of, 89–90
El Cid, Falangist ideology and, 108–109
El Desaster (The Disaster), 2–5, 40–44,
    223
Elimelech, Menachem, 204
El Imparcial (newspaper), 77, 79–80
Eliot, T. S., 39
Elites of Spain. *See also Latifundistas*
    autarchic hydrological development
        and, 135–139
    Corps of Engineers and, 116
    democratization and, 165–171
    Falangist ideology and, 108

Franco regime and, 99, 104, 116,
    123–127
historical-geographical materialism
    and, 25
hydro-justice and, 34
hydrological divisions and, 85–93
hydro-social cycle and emergence of
    new elites, 14–16
integrated water system proposals and,
    174–176
loss of empire and, 3
Marshall Plan and, 129–132
National Hydrological Plan of 2000
    and, 189–190
postcolonial shock and, 40–44
Primo de Rivera dictatorship and,
    68–74
regeneracionismo and, 45–64
rescaled networks of interest by, 139,
    142–149
state institutions and, 74–76, 79–81
El Sexenio Recoluvionario (Spain,
    1868–1874), 40–44
El Sol (newspaper), 89
"El victimismo" ideology, regional
    autonomy and, 216–217
Energy development, autarchic control
    of, 136–139
Engels, Friedrich, 26–27
Entrepeñas/Buendia dam complex, 149
Environmental politics
    absence of water in, 227–230
    desalination technology and,
        194–221
    hydro-justice and, 33–35
    integrated water system proposals and,
        176–178
    mercantilization of water policy and,
        206–207
    modernization and globalization and,
        207–213
    neoliberal policies and, 182–185
    water management and, 7–9

Equity, hydro-justice and, 33–35
Estudios Geográficos, 121
European Economic Community
 draft NHP proposals and, 177–178
 Spain and, 167, 181, 218–219
European Environment Agency, 204
European integration, Spanish water politics and, 180–185
European Parliament, anti-dam protests at, 163
European socialism, decline of, 180–185
European Union, 17–18
 "Blue March" protest at headquarters of, 189, 189f
 desalination technology and, 200
 neoliberal ideology and, 183–185
 river basin governance and, 32
 Water Framework Directive, 189, 218–219
Euskadi Ta Askatasuna (Basque Homeland and Freedom) (ETA)
 Madrid terrorist attacks of 2004 and, 191
 regional autonomy demands of, 167, 244n3

Falange Española party, 89
 Corps of Engineers and, 116–119
 Franco's water policies and, 108–109
 patriotic geography and, 121–123
Falange Española Tradicionalista y de las J.O.N.S., 151
Farley, James, 144
Fascist (Falangist) ideology
 hydraulic politics and, 65–66, 89, 108–109
 U.S. views on, 147–148
Fernández de la Mora, Gonzalo, 153
Feudal water principles, history in Spain of, 70–74
First National Water Plan (Spain), 82–85, 83t
Fitzcarraldo (Herzog), 1–2

Flecha, Ricardo, 159
Fordist development model
 democratization and evolution of, 166
 interregional water management and, 180–185
 Spanish emigration and, 137
Foucault, Michel, biopolitical theory of, 4, 32–33
France, Franco regime and, 135–139, 144
Franco, Generalissimo Francisco Bahamonde, 16, 57, 98–99
 Corps of Engineers and, 116–119, 120t
 dam construction under, 100–101
 democratization and legacy of, 168–171, 172–176
 fascist (Falangist) ideology and, 65, 108–109
 forced labor by prisoners under, 137–139, 140t–141t
 history of regime, 4–5
 hydraulic infrastructure during regime of, 14
 hydro-electrification projects and, 111–116, 113f
 modernization and role of, 16–18
 national development and autarchy of, 135–139
 politics of water and regime of, 101–108, 102f
 propaganda and hydraulic projects during regime of, 123–127
 Tajo-Segura transfer system and, 150–156

García Berlanga, Luis, 129–132
García-Diego de la Huerga, Tomás, 117–119
Gasset, Rafael, 77, 79–81, 83–84, 87
Gasson, Christopher, 207
Genealogy of narrative, 19–24
General Agreement on Tariffs and Trade (GATT), 177

Index

Generales, Cortes, 1
Generalidad de Cataluña, water management and, 95
"Generation of '98," 49–50
  Franco's dismissal of, 99
  modernization of Spain and, 42–44
  regenerationism and, 53–57
Geography
  Franco regime's use of, 121–123
  hydroelectrification projects and, 114–116
  national development and integration of, 135–139
  postcolonial Spain and importance of, 41–44
  as propaganda symbol, 124–127
  in regenerationist literature, 54–57
  space and land issues and, 68–70
  Spanish modernization and re-engineering of, 44–51
Germany, hydro-social interventions and modernization in, 31–33
Globalization
  climate change and, 204–206
  desalination technology and, 207–213
  hydro-social politics and, 17–18
  Spanish water policies and role of, 181–185
Gomez De Pablos, M., 100–101
Gonzáles, Matteu, 71
González, Felipe, 167, 181
González Bernáldez, 200–201
González Bueno, Petro, 118
Griffis, Stanto, 145
Guadalquivir River Basin Authority, 137–139, 172
Guattari, Félix, 19
Guernica, bombing of, 142
Guerra del agua (water war), regional autonomy and, 214–217
Guerra del Rio, Rafael, 96
Gurney, Chan, 144

Hacia otra España (de Maeztu), 43
Haraway, Donna, 8–9
Harvey, David, 7, 23–24
Herzog, Werner, 1–2
Heterogeneous assemblages, of hydro-social cycle, 35–37
Hidroeléctrica Moncabril, 157
High council for Scientific Research (HCSR), 121
Hilton, Contrad, 145
Hilton Hotels, 145
Historia de España (Pérez de la Dehesa), 51
Historical-geographical materialism
  nature and, 20
  socio-ecological perspective, 25–28
"Hydraulic desire" motif, in regenerationist literature, 54–57
Hydraulic engineering
  emergence of despotism in, 93–94
  failure of early twentieth-century Spanish modernization of, 67–98
  under Franco regime, 106–108
  Franco's propaganda concerning, 123–127, 126f
  Generation of '98 and, 53–57
  historical legacy in Spain of, 64–66
  history in Spain of, 4–5
  mercantilization of water and decline of, 206–207
  metabolic role of state in, 76–85
  modernization in Spain and, 46–51
  post-Franco interventions, 163–190
  purification and transformation of nature and, 57–64
  regeneracionismo and modernization of, 39–66
  Spanish Civil War impact on, 94–98
  state role in, 74–76
"Hydraulic structuralism"
  democratization and, 167–171
  protests against, 165–166

state involvement in water projects and, 77–85
Hydroelectrification
  autarchic control of, 136–139
  Franco's propaganda concerning, 123–127, 126f
  Spanish Civil War and, 111–116, 113f
Hydro-infrastructure planning, 32–33
  U.S. aid and, 146–149
Hydro-justice
  exploration of, 33–35
  Franco regime and, 101–108
  future issues for, 228–230
"Hydrological Disequilibrium," Pardo's concept of, 97, 165–166
Hydrological Service and Divisions (Spain)
  creation of, 81–85, 87–88
  political struggles within, 85–88
Hydro-politics
  climate change and, 202–206
  Cold War politics and, 145–149
  desalination technology and, 194–221
  early twentieth-century emergence of, 51–53, 60–66, 80, 82–85
  forced labor for dam construction and, 137–139
  future issues in, 223–230
  interregional water wars and, 179–185
  nationalist ideology and, 133–135
  neoliberal policies and, 182–185
  regional autonomy in post-democratic era and, 179–185
  Tajo-Segura transfer system and, 150–156
Hydro-social cycle
  aridity and management of, 200–202
  Cold War politics and, 146–149
  democratization and, 165–171
  desalination technology and, 192–221

early twentieth-century modernization attempts and, 67–68
  under Franco regime, 132–135
  legacy of Franco's policies in, 157, 158f, 159
  metabolic circulation of, 35–37
  modernization and, 7–9, 19–37
  neoliberal policies and, 182–185
  politicization of matter and, 28–30
  production of nature, 20–24
  in regenerationist literature, 54–57
  structure of, 14–18
  Tajo-Segura transfer system and, 150–156
  transformation of matter through, 28–30
Hydro-state, Primo de Rivera dictatorship and emergence of, 88–93

Iberian Peninsula, map of, 10f
Imaginary
  of capitalist accumulation, 75–76
  desalination technology and role of, 200–202
  politics and role of, 15
  in postcolonial Spain, 41–44
  Spanish Civil War impact on hydraulic policy and, 94–98
Import substitution industrialization, autarchic mobilization of, 135–139
India, nation-state formation and modernization in, 31–33
Industrialization
  import substitution and, 135–139
  Spanish social conflict and role, 44–51
Institución Libre de Ensenanza (Free Institute of Education), 49, 233n15
Institute for Foreign Trade (ICEX) Spain, 207–213
Institute of Geography, 121
Instituto Nacional de Colonización (INC), 110–111

Index

Integrated resource development
  autarchic mobilization of, 135–139
  democratization and integrated water management, 171–176
  Spanish river basin management and, 91–93
  Tajo-Segura transfer system and, 150–156
Integrated System of National Hydraulic Equilibrium (Sistema Integrado Hidráulico Nacional [SIEHNA]), 106–107, 173–176
Intergovernmental Panel on Climate Change (IPCC), 203
International Financial Corporation, 146
International Monetary Fund, 146
Irrigation projects
  autarchic control of hydraulic development and, 136–139
  democratization and policies involving, 168–171
  Franco regime and, 109–111, 132
  Franco's propaganda concerning, 123–127, 126f
  National Irrigation Plan, 177
  national water plans and, 82–85
  Spanish water laws and, 71–74
  state role in, 74–76
Italy
  Franco's alliance with fascism in, 129, 136, 142, 146
  hydro-social interventions and modernization in, 31–33
Itoiz Dam project, protests against, 163, 164f, 165–166

John Paul II (Pope), 6, 189–190
Junta Consultiva de las Obras Públicas, 60

Kaika, Maria, 15
Kennedy, John F., 191
Keogh, Eugene, 144
Kinski, Klaus, 1

Labor
  autarchic control of hydraulic development and supply of, 137–139
  transformation into power of, 26–28
Labor Code (decree of 1938), 109
Lafuente, Isaías, 139
La invertebración of Spain, 43, 232n5
La Movida (the movement/scene), 167
Land redistribution
  Spanish Civil War and, 109–111
  state-sponsored water policies and, 76
  water politics and, 68–70
Latifundistas
  hydraulic despotism and, 93–94
  hydraulic engineering and declining power of, 64–66
  land distribution and water politics and, 69–70
  Spanish Civil War and, 109–111
  Spanish modernization and role of, 44–51, 81–85
  trade liberalization and decline of, 74
Latour, Bruno, 8–9, 20–22
Law of 1866, Spanish water policy under, 71–74
Law on University Education (1943), 122–123
Lefebvre, Henri, 2–240
Lequerica, José Frlix, 144
Levant region
  hydraulic politics and, 82, 97, 124, 146
  Tajo-Segura transfer system and, 150–156
Leyenda de Oro (legend of Golden Spain), 41, 53–57
Liberal revolution in Spain (1811–1873), 71
Libro Blanco del Agua en España (The White Book on Spanish Water) (Ministerio de Medio Ambiente 1998), 186
Lion of Graus. See Costa, Joaquin

Lisbon Treaty of 2007, 218
London Eye, protests at, 163

Maastricht Treaty of 1992, 183, 218
Machada, Antonio, 53–54
Maciás Picavea, Ricardo, 49, 52–54, 56–57
Macquarie investment fund, 205
Madrid
  terrorist attacks of 2004 in, 191
  water provision of 1858 in, 73
Maeztu, Ramiro de, 43, 54–55, 80
Mallada, Lucas, 49, 54
Malthus, 202–206
Maluquer de Motes, 71
Manchesterismo, 81
Mancommunidades (river basin authorities), 95
Mancomunidad Hidrográfica del Ebro, 95
Manifest destiny doctrine, U.S. policy toward Spain and, 243n5
March, Hug, 200
Marcuello, José Ramón, 93–94
Margall, Pasqual, 181
Maria Christina (Queen-Regent), 80
Market forces
  autarchic control of hydraulic development and, 136–139
  democratization and, 168–171
  neoliberal policies and, 182–185
  state-framed water development and, 74–76, 80–85
Marshall Plan, exclusion of Spain from, 129–132, 146–147
Martínez Gil, F. J., 107, 178
Marx, Karl, 25–28, 235n1
Matter
  desalination and politics of, 200–202
  political transformation of, 28–30
McCarran, Pat, 144
McCarthy, Joe, 144

Media coverage of water policies
  during Franco regime, 104, 106–108
  Franco's propaganda instruments and, 123–127, 126f
  U.S.-Spanish relations and, 147–149
  Vega de Tera dam collapse and, 157
Medina, José Antonio, 208
Membrane technology, desalination and, 209–212, 209f, 212f
Mendiluce, Martín, 172
Mendoza, Gómez, 48
Mendoza Gimeno, José Luis, 118–119
Mercantilization of water policy, 183–185
  desalination mythologies and, 200, 206–207
Metabolic circulation
  historical-geological transformation and, 26–28
  of hydro-social cycle, 35–37
  state role in hydraulic engineering and, 76–85
Military Directory, creation of, 89
Millennium Wheel, protests at, 163
Minifundios, Spanish modernization and role of, 44–51
Modernity/modernization
  contested process of, 7–8
  desalination technology and, 207–213
  early twentieth-century acceleration of, 67–68
  Franco's "networks of interest" and, 134–135
  mobilization of seawater management and, 191–221
  "networks of interest" and, 108
  post-Franco reimagining of, 163–190
  regeneracionismo and hydraulic modernization, 39–66
  role of imaginary in, 15
  socio-natural networks and, 20–224
  Spanish transition to, 2–5

Index

state role in Spain's move to, 74–76
water's role in, 1–5, 51–53, 99–101
Morales Amores, Antonio, 60, 79
Movies, water in, 1–5
Mussolini, Benito, 90

Narbona Ruiz, Cristina, 191–192, 206–207
Narcissism, regional autonomy and, 216–217
Narmada Dam Project (India), 163
"National-Catholic" doctrine, Franco's patriotic geography and, 122–123
National City Bank, 145
National Council for Public Works, 136
National Falangist Movement, 148
National hydraulic development
  democratization and disintegration of, 179–185
  Franco regime's water policies and, 103–108
  integrated system of, 171–176
  protests against, 163, 164f, 165–166
  Spanish proposals for, 77–85
National Hydrological Plan (2000), 151, 186–187, 188f
  approval of parts of in 2001, 191–192
  protests against, 187, 189–190, 189f
National Irrigation Plan, 177, 185–186
Nationalist ideology
  democratization and re-emergence of, 167–171
  Falange party and, 108–109
  Franco's patriotic geography and, 122–123
  "hydraulic nationalism" and, 216–217
  modernization and, 16
  propaganda and hydraulic projects and, 123–127
  regeneracionismo and, 42–44
  regional river basin management marginalization and, 133–135
National Park policy (Spain), 59–64

National Plan for Hydraulic Works (1933), 96–98, 133, 151
National Water Council (NWC), 176–177, 206–207
Nation-state formation. *See also* State-centered water policies
  early twentieth-century hydro-social politics and, 67–68
  geography and, 50–51
  hydro-social narrative and, 30–33
  Primo de Rivera dictatorship and emergence of hydro-state, 88–93
  socio-environmental engineering and, 74–76
  Spanish water rights and, 71–74
NATO (North Atlantic Treaty Organization), Spain's membership in, 181
Nature
  democratization and role of, 168–171
  desalination mythologies and, 200
  hydro-politics and, 228–230
  hydro-social justice and, 33–35
  purification and transformation of, hydraulic engineering and, 57–64
  Smith's "production of," 20–24, 47
Nazi Germany, Franco and, 129, 136, 142–143, 146
Neoliberal reform
  democratization and, 167
  mercantilization of water policy and, 206–207
  water system management and impact of, 179–185
"Networks of interest"
  desalination technology and, 199–200
  Franco's modernization projects and, 134–135
  hydroelectrification projects and, 114–116
  Spanish modernization an, 108
  U.S. aid and rearrangement of, 146–149

NO-DO *(Noticiario Español Cinematográfico)*, 157
propaganda projects by, 123–127, 126f

Obras Completas (de Unamuno), 53
O'Konski, Alvin, 144
Oligarquía y Caciquismo (Costa), 47
Opus Dei, technocrats from, 147
Organic Machine, The (White), 8–9
Organisation for Economic Co-operation and Development (OECD), Spain's admission to, 147
Oriental Despotism (Wittfogel), 30–33
Ortega y Gasset, José, 55, 89
Ortí, Alfonso, 50–51, 53, 75
Ottoman Empire, 44

Paco el Rana (Frankie the Frog) (Franco's nickname), 101–108, 124–127, 159
"Pact of Madrid," 146, 243n5
Palancar Penello, Mariano, 172
Para la Nueva Política (In Defense of the New Politics) (Gasset), 79–80
Pardo, Manuel Lorenzo, 82–83, 90–98, 136, 150–151, 238n24
Partido Popular (Popular Party, PP), 155, 177–178, 182
  desalination technology and, 195–199, 216–217
  Madrid terrorist attacks of 2004 and, 191
  National Hydrological Plan of 2000 and, 189–190
Partido Republicano Radical, 96
Patriotism
  hydraulic politics and, 80
  Spanish Civil War and geography of, 121–123
  Spanish geography linked to, 49–50
Pearson, Drew, 144
Peasant movement
  land redistribution and, 76
  water politics and, 68–70, 109–111

Peña Boeuf, Alfonso, 118, 133
People's Republic of China, nation-state formation and modernization in, 31–33
Pérez de la Dehesa, Rafael, 45, 51
Pertinaz sequía (perennial drought), Franco's propaganda concerning, 114, 135
Philippines, Spanish colonization of, 40–44
Phillip, William, 204
Phylloxera infestation, devastation of Spanish agriculture and, 44
Picasso, Pablo, 142
Plan Badajoz, 111–116
Plan de Obras Hidráulicas (POH) (Plan for Hydraulic Works [PHW]), 81–85
Plan General de Obras Públicas (General Plan for Public Works), 133, 136
Platform for the Defense of the Ebro *(Plataforma en Defensa del Ebro)*, 187, 189–190, 189f, 214–217
Plato, 34
Political prisoners, as forced labor source, 137–139, 140t–141t
Politics
  caciquismo in Spain of, 41–44
  Costa's hydraulic politics and, 51–53
  Generation of '98 and, 54
  hydro-scales and, 30–33
  land distribution and, 68–70
  socio-environmental transformation in, 19–24
  state role in hydraulic engineering and, 74–76
  transformation of matter through, 28–30
  water and, 1, 7–9
Popular Front movement, hydraulic politics and, 94–98, 238n24
Postcolonialism
  shock in Spain of, 40–44
  state debt burden in era of, 81–85

Index

Poverty, Spanish geography linked to, 48–51
Prieto, Indalecio, 96–98
Primo de Rivera, General Miguel, 11, 68, 85, 88, 108
 emergence of the hydro-state (1923–1930) and, 88–93, 133
 hydraulic despotism and, 93–94
Primo de Rivera, José Antonio, 89, 108–109
Private water management initiatives
 failure in Spain of, 70–74
 hydro-electrification projects and, 111–116, 113f
 neoliberal ideology and rise of, 183–185
Programa A.G.U.A., 192, 195–221, 197t, 198t, 206–207, 219–221
Propaganda, Franco's hydraulic projects and, 123–127
"Pseudo-regenerationist" politics, 81
Public domain laws, Spanish water rights and, 71–74
Public water projects
 evolution in Spain of, 70–74, 81–85
 Franco regime and, 118–119
 neoliberal policies and decline of, 183–185
 Primo de Rivera dictatorship and, 89–93
Puerto Rico, Spanish colonization of, 40–44

Quasi-object theory (Latour), 8–9

Rajoy, Mariano, 219
Ramón y Cajal, Santiago, 43–44
Reed, Ralph, 145
Regeneracionismo
 democratization and legacy of, 168–171
 Franco and ideology of, 103–108
 "Generation of '98" and, 53–57
 historical legacy of, 64–66
 hydraulic engineering vs.reforestation debate and, 57–64
 hydraulic metabolism and, 76–85
 integrated water system proposals and legacy of, 175–176
 land distribution and water politics and, 68–70
 modernization in Spain and, 44–51
 postcolonial emergence of, 42–44
 Primo de Rivera dictatorship and, 89–93
 rural ideal in, 74–76
 water scarcity and, 51
Regional water management
 democratization and re-emergence of, 165–171, 179–185
 desalination technology and, 195–221, 214–217
 state-centered water policies *vs.*, 133–135
Republic (Plato), 34
Republic period in Spain, hydraulic infrastructure during, 14
Reservoir capacity
 under Franco regime, 132
 U.S. aid and expansion of, 149
Restoration period in Spain, 40–45
 hydrological divisions during, 87–88
Reverse osmosis technology, desalination and, 208–209, 211–213
Revista de la Confederación Sindical Hidrografica del Ebro, 93–94
Revista de Obras Públicas (ROP), 60, 78–79, 94
 on desalination technology, 193
 Franco regime and, 116–119, 120t
 on hydrological divisions, 87
 integrated water system proposals and, 171, 175
Ribadelago de Francoo, construction of, 157–161, 160f
Rice cultivation, in Spain, 5

River basin authorities (RBAs),
  establishment of, 179–185
River basin management
  annual river flow statistics, 203t
  climate change and, 203–206
  dam construction and, 115–116, 115t
  democratization and policies
    involving, 168–171
  draft NHP proposals for integration of,
    173–176
  under Franco, 100–101, 133–135
  freshwater resource availability and
    demand and, 170t
  hydraulic engineering in Spain and,
    59–64
  hydro-social cycle and, 32–33
  map of Spanish authorities and
    hydrogeographic confederations,
    62f
  National Hydrological Plan (2000)
    and, 186–187, 188f
  Primo de Rivera dictatorship and,
    91–93
  regional autonomy and, 179–185
  sociedades estatales (state companies)
    as replacement for, 214–217
  Spanish Civil War impact on, 94–98
  Spanish politics and, 85–88
  U.S. aid and rearrangement of, 149
Rodríguez, S., 123–124
Roman Catholic Church
  American Catholic support for Franco
    and, 144–145, 147–149, 243n4
  Franco regime and, 109, 122–123, 143
  Spanish elites and, 3, 76
  water protests and, 189–190
Roosevelt, Franklin Delano, 32–33, 92,
  142, 144, 243n4
Ruiz, José Martínez (Azorín), 42, 49–50,
  54, 80
Ruiz, Juan Manuel, 168
Rural ideal, agricultural policy and,
  74–76

Russia. *See also* Soviet Union
  wheat imports to Spain from, 44

Sábado Gráfico magazine, 157
Sánchez-Albornoz, Nicolás, 110–111
Saudi Arabia, desalination capacity in,
  209
Scalar strategies
  desalination technology and,
    193–195, 200, 214–217
  Franco's legacy and, 159–161
  Franco's use of, 132
  hydro-scales and politics, 30–33
  Spanish Civil War impact on hydraulic
    policy and, 94–98
  U.S. aid and, 148–149
  Water Framework Directive, 189,
    218–219
Schaefer, William Donald, 181
Schmidt, G., 196
Schnaiberg, Alan, 26
Second National Hydrological Plan,
  Draft Proposal of, 106–107, 234n 22
Second Republic (Spain), 94–98
Segura River Basin Authority, 91–93
Self-rule, land distribution politics and,
  68–70
Self-sufficiency, land distribution
  politics and, 68–70
Sherman, Forrest P., 145
Siemens company, 205
Silva Muñoz, Federico, 152
Silvela, Francisco, 80–81
Simpson, James, 111
Smith, Neil, 20, 47
Smith, T. C., 61, 63
Socialization of matter, 28–30
Social power relations
  hydro-politics and, 223–230
  metabolic circulation and, 36–37
Sociedades estatales (state companies),
  183–185, 196, 214–217
Sociedad Geográfica de Madrid, 48–51

Socioeconomic conditions
  Falangist ideology and, 108–109
  forced labor by prisoners under,
    137–139, 140t–141t
  water and, 1–5
Socio-environmental transformation
  Franco regime's water policies and,
    103–108
  historical-geographical materialism
    and, 25–28
  politics and, 19–24
Socio-nature
  Franco's legacy and, 159–161
  hydraulic engineering and production
    of, 57–64
  production of, 20–24
Sociotechnical systems, formation of,
  32–33
"Solid@rios con Itoiz," 163, 164f,
  165–166
South Africa, nation-state formation
  and modernization in, 31–33
Sovereign development, land
  distribution politics and, 68–70
Soviet Union. *See also* Russia
  nation-state formation and
    modernization in, 31–33
Spain
  Europeanization of, 74–76
  evolution of dam construction in,
    11–14, 12f–13f
  hydraulic modernization in, 39–66
  hydro-electrification in, 111–116
  hydro-social landscape in, 9–14
  imperial defeat of, 15
  nation-state formation and
    modernization in, 31–33
  political-ecological history of, 2–5,
    11–14, 12f
  postcolonial shock in, 40–44
  rice cultivation in, 5
  U.S. military bases in, 146–149
  water basin management in, 32–33

Spanish and International Committee
  for Large Dams, 119
Spanish Civil War
  alliance with Nazi Germany and
    Fascist Italy during, 142
  Brenan's account of, 127
  Corps of Engineers and, 116–119,
    120t
  Falangist ideology and, 108–109
  hydraulic policies and, 94–98
  hydro-electrification and, 111–116
  latifundistas and, 109–111
  modernization and role of, 15–16,
    99–101, 127–128
  patriotic geography and, 121–123
  regenerationist agenda and, 50
  U.S. views on, 147–148, 243n4
Spanish Desalination and Water Reuse
  Association (AEDyR), 207–213
Spanish Labyrinth, The (Brenan), 127
Spanish Socialist Workers Party (Partido
  Socialista Obrero Español, PSOE),
  167, 177–178, 181–182
  defeat of, 219–221
  desalination technology and,
    195–200, 216–217
  Programa A.G.U.A. and, 206–207
Spellman, Francis Cardinal
  (Archbishop), 144
Stabilization Plan of 1959, 147
State-centered water policies. *See also*
  Nation-state formation
  democratization and changes in,
    167–171, 180–185
  early twentieth-century hydro-social
    politics and, 67–68
  Franco regime and, 103–108
  geography and, 50–51
  hydraulic despotism and, 93–94
  hydraulic metabolism and, 76–85
  hydro-social narrative and, 30–33
  neoliberal ideology and decline of,
    182–185

regional water management and
dominance of, 133–135
socio-environmental engineering and,
74–76
water rights and, 71–74
Structural and Cohesion Funds
(European Union), 183
Styles Bridges, Henry, 129, 148
Suárez Fernández, L., 1590151
Syndicalist Organization and Action,
118–119

Tajo River Basin Authority, 91–93
Tajo-Segura transfer system, 106–107,
132, 150–156, 153f, 154f, 155f, 156f
regional autonomy and, 215–217
Techno-natural water management
democratization and, 168–171
desalination technology and, 192–195
Franco regime and, 101–108
modernization and globalization and,
207–213
narrative of, 19–24
U.S. aid and, 146–149
Tejero, Lieutenant-Colonel Antonio, 167
Tennessee Valley Authority (TVA),
32–33, 92
Territoriality
Franco's national territorial
integration and, 100–101
scalar networks and, 30–33
Thames Water, 205
"Third Way" politics
democratization and, 167
Spanish water policies and, 181–185
Thorning, Joseph F., 144
Tierra de Campos (Macías Picavea), 54,
56–57
Torres Padilla, C., 172
Tourism in Spain, 5–6
Cold War politics and promotion of,
145–146
democratization and, 168–171

Trade liberalization
autarchic control of hydraulic
development and, 136–139
Spanish agriculture and, 74–76
Trans World Airlines, 145
Truman, Harry S., 144–145
Tuñón de Lara, M., 81
Turkey, nation-state formation and
modernization in, 31–33

Ullastres, Alberto, 147
Unamuno, Miguel de, 53–54, 80
Unión Fenosa, 157
Union General de Trabajadores, 116
Union of the Democratic Centre (*Unión
de Centro Democrático* [UDC]), 155
United Arab Emirates (UAE),
desalination capacity in, 209
United Kingdom
desalination in, 205
Franco regime and, 135–139, 144
United Nations
Security Council subcommittee on
"the Spanish Question," 143
Spanish membership in, 146–149
United Nations Educational, Scientific
and Cultural Organization
(UNESCO), 146
"Water for All" slogan of, 215–217,
215f
United States
Cold War and Spanish relations with,
142–149
defeat of Spain by, 40–44
democratization of Spain supported
by, 167–168
desalination capacity in, 209
Franco regime and, 135–139, 142–149
hydro-social interventions and
national identity in, 31–33
Marshall Plan assistance from,
129–132
wheat imports to Spain from, 44

Index

"Un Profeta Político" (Varela), 39
Urban migration
   autarchic control of hydraulic development and, 136–139
   in postcolonial Spain, 40–44

Varela, Javier, 39, 65
Vatican, anti-dam protests at, 163
Vega de Tera dam collapse, 157, 158f, 159
Veolia company, 205
Vertebración, 175, 244n6
Vibrant Matter (Bennett), 200–202

Walters, Vernon, 243n5
Waste Land, The (Eliot), 39
Water banks, neoliberal establishment of, 184–185
Water Framework Directive (EU), 189, 218–219
Water Law amendments (1999), 184–185, 214–217
Water Law of 1879, 72, 86
Water Law of 1985, 72, 171, 179, 182, 185–186
Water markets, neoliberal establishment of, 184–185
Water Pact (*Pacto del Agua*), 179–180
Water payment system, evolution in Spain of, 72–74
Water rights, history in Spain of, 70–74
Water transfer and diversion projects. *See also* Tajo-Segura transfer system
   democratization and integrated water systems, 171–176, 174f
   desalination technology and, 191–221
   National Hydrological Plan (2000) and, 186–187, 188f
   Spanish protests over, 6, 163–164
Water Wars in Spain, 6

Western Alliance
   Franco and, 16, 100–101, 135–139, 143–149
   Spain's admission into, 148–149
White, Richard, 8–9
Wittfogel, Karl, 30–33
World Water Forum, 163
World Wildlife Fund, 177–178
Worster, Donald, 9

Zapatero, José Luis, Rodriguez, 191–192, 195–196, 219
Žižek, Slavoj, 19, 228–230

Urban and Industrial Environments

Series editor: Robert Gottlieb, Henry R. Luce Professor of Urban and Environmental Policy, Occidental College

Maureen Smith, *The U.S. Paper Industry and Sustainable Production: An Argument for Restructuring*

Keith Pezzoli, *Human Settlements and Planning for Ecological Sustainability: The Case of Mexico City*

Sarah Hammond Creighton, *Greening the Ivory Tower: Improving the Environmental Track Record of Universities, Colleges, and Other Institutions*

Jan Mazurek, *Making Microchips: Policy, Globalization, and Economic Restructuring in the Semiconductor Industry*

William A. Shutkin, *The Land That Could Be: Environmentalism and Democracy in the Twenty-First Century*

Richard Hofrichter, ed., *Reclaiming the Environmental Debate: The Politics of Health in a Toxic Culture*

Robert Gottlieb, *Environmentalism Unbound: Exploring New Pathways for Change*

Kenneth Geiser, *Materials Matter: Toward a Sustainable Materials Policy*

Thomas D. Beamish, *Silent Spill: The Organization of an Industrial Crisis*

Matthew Gandy, *Concrete and Clay: Reworking Nature in New York City*

David Naguib Pellow, *Garbage Wars: The Struggle for Environmental Justice in Chicago*

Julian Agyeman, Robert D. Bullard, and Bob Evans, eds., *Just Sustainabilities: Development in an Unequal World*

Barbara L. Allen, *Uneasy Alchemy: Citizens and Experts in Louisiana's Chemical Corridor Disputes*

Dara O'Rourke, *Community-Driven Regulation: Balancing Development and the Environment in Vietnam*

Brian K. Obach, *Labor and the Environmental Movement: The Quest for Common Ground*

Peggy F. Barlett and Geoffrey W. Chase, eds., *Sustainability on Campus: Stories and Strategies for Change*

Steve Lerner, *Diamond: A Struggle for Environmental Justice in Louisiana's Chemical Corridor*

Jason Corburn, *Street Science: Community Knowledge and Environmental Health Justice*

Peggy F. Barlett, ed., *Urban Place: Reconnecting with the Natural World*

David Naguib Pellow and Robert J. Brulle, eds., *Power, Justice, and the Environment: A Critical Appraisal of the Environmental Justice Movement*

Eran Ben-Joseph, *The Code of the City: Standards and the Hidden Language of Place Making*

Nancy J. Myers and Carolyn Raffensperger, eds., *Precautionary Tools for Reshaping Environmental Policy*

Kelly Sims Gallagher, *China Shifts Gears: Automakers, Oil, Pollution, and Development*

Kerry H. Whiteside, *Precautionary Politics: Principle and Practice in Confronting Environmental Risk*

Ronald Sandler and Phaedra C. Pezzullo, eds., *Environmental Justice and Environmentalism: The Social Justice Challenge to the Environmental Movement*

Julie Sze, *Noxious New York: The Racial Politics of Urban Health and Environmental Justice*

Robert D. Bullard, ed., *Growing Smarter: Achieving Livable Communities, Environmental Justice, and Regional Equity*

Ann Rappaport and Sarah Hammond Creighton, *Degrees That Matter: Climate Change and the University*

Michael Egan, *Barry Commoner and the Science of Survival: The Remaking of American Environmentalism*

David J. Hess, *Alternative Pathways in Science and Industry: Activism, Innovation, and the Environment in an Era of Globalization*

Peter F. Cannavò, *The Working Landscape: Founding, Preservation, and the Politics of Place*

Paul Stanton Kibel, ed., *Rivertown: Rethinking Urban Rivers*

Kevin P. Gallagher and Lyuba Zarsky, *The Enclave Economy: Foreign Investment and Sustainable Development in Mexico's Silicon Valley*

David N. Pellow, *Resisting Global Toxics: Transnational Movements for Environmental Justice*

Robert Gottlieb, *Reinventing Los Angeles: Nature and Community in the Global City*

David V. Carruthers, ed., *Environmental Justice in Latin America: Problems, Promise, and Practice*

Tom Angotti, *New York for Sale: Community Planning Confronts Global Real Estate*

Paloma Pavel, ed., *Breakthrough Communities: Sustainability and Justice in the Next American Metropolis*

Anastasia Loukaitou-Sideris and Renia Ehrenfeucht, *Sidewalks: Conflict and Negotiation over Public Space*

David J. Hess, *Localist Movements in a Global Economy: Sustainability, Justice, and Urban Development in the United States*

Julian Agyeman and Yelena Ogneva-Himmelberger, eds., *Environmental Justice and Sustainability in the Former Soviet Union*

Jason Corburn, *Toward the Healthy City: People, Places, and the Politics of Urban Planning*

JoAnn Carmin and Julian Agyeman, eds., *Environmental Inequalities Beyond Borders: Local Perspectives on Global Injustices*

Louise Mozingo, *Pastoral Capitalism: A History of Suburban Corporate Landscapes*

Gwen Ottinger and Benjamin Cohen, eds., *Technoscience and Environmental Justice: Expert Cultures in a Grassroots Movement*

Samantha MacBride, *Recycling Reconsidered: The Present Failure and Future Promise of Environmental Action in the United States*

Andrew Karvonen, *Politics of Urban Runoff: Nature, Technology, and the Sustainable City*

Daniel Schneider, *Hybrid Nature: Sewage Treatment and the Contradictions of the Industrial Ecosystem*

Catherine Tumber, *Small, Gritty, and Green: The Promise of America's Smaller Industrial Cities in a Low-Carbon World*

Sam Bass Warner and Andrew H. Whittemore, *American Urban Form: A Representative History*

John Pucher and Ralph Buehler, eds., *City Cycling*

Stephanie Foote and Elizabeth Mazzolini, eds., *Histories of the Dustheap: Waste, Material Cultures, Social Justice*

David J. Hess, *Good Green Jobs in a Global Economy: Making and Keeping New Industries in the United States*

Joseph F. C. DiMento and Clifford Ellis, *Changing Lanes: Visions and Histories of Urban Freeways*

Joanna Robinson, *Contested Water: The Struggle Against Water Privatization in the United States and Canada*

William B. Meyer, *The Environmental Advantages of Cities: Countering Commonsense Antiurbanism*

Rebecca L. Henn and Andrew J. Hoffman, eds., *Constructing Green: The Social Structures of Sustainability*

Peggy F. Barlett and Geoffrey W. Chase, eds., *Sustainability in Higher Education: Stories and Strategies for Transformation*

Isabelle Anguelovski, *Neighborhood as Refuge: Community Reconstruction, Place-Remaking, and Environmental Justice in the City*

Marianne Krasny and Keith Tidball, *Civic Ecology: Adaptation and Transformation from the Ground Up*

Erik Swyngedouw, *Liquid Power: Water and Contested Modernities in Spain, 1898–2010*